Hazardous Materials Compliance for Public Research Organizations

A Case Study, Second Edition

Hazardous Materials Compliance for Public Research Organizations

A Case Study, Second Edition

Nicolas A. Valcik

CRC Press is an imprint of the
Taylor & Francis Group, an **informa** business

CRC Press
Taylor & Francis Group
6000 Broken Sound Parkway NW, Suite 300
Boca Raton, FL 33487-2742

Printed on acid-free paper
Version Date: 20121220

International Standard Book Number: 978-1-4665-0946-7 (Hardback)

Visit the Taylor & Francis Web site at
http://www.taylorandfrancis.com

and the CRC Press Web site at
http://www.crcpress.com

To my wife, Kristi

Contents

List of Figures

List of Tables

List of Photographs

Foreword

Over the past fifty years, colleges and universities have experienced manifold increases in the number of scientific studies and the complexity of the research. The academic researcher of today can now call upon research techniques to investigate a range of scientific phenomena that were simply unavailable to colleagues even just one or two decades ago. In particular, there has been a substantial increase in research activities that involve radiological, biological, or chemical agents and such studies often involve either dangerous or potentially hazardous specimens as well as the waste products of the research. Although it might be expected that there are uniform guidelines for the conduct of research with dangerous materials, or that compliance with such guidelines is ensured through strict monitoring and auditing by federal agencies, the reality of the situation is far from definitive. Essentially, guidelines and regulations are still evolving and compliance seems to be a matter of degree rather than an absolute.

In this text, Nicolas Valcik employs a case study approach of one university's policies and practices with regard to the procurement, use, storage, and disposal of biological hazardous materials (HAZMAT). For purposes of the study, he treats biological HAZMAT as certain select agents, controlled substances that induce biological ailments, animals, plants, blood-borne pathogens, human waste by-products, tissue samples, and other waste produced from biological research experimentation. By adopting a case study of a single institution, Dr. Valcik provides an organizational theory approach to the examination of how changing university structures and administrative policies and procedures function within a changing regulatory environment, thus producing a set of generalizable observations, conclusions, and recommendations.

There are a number of aspects to this research that strengthen its value for public policy, but there are three features in particular that are worth comment. First, Dr. Valcik provides a comprehensive historical analysis of the various regulatory schemes that have evolved to ensure the safety of research using biological HAZMAT. In this context, he highlights how recent legislation such as the Patriot Act of 2001 and the Bioterrorism Preparedness and Response Act of 2002 have focused more attention and scrutiny on how research universities are expected to comply with new federal guidelines. Second, the study employs a set of well-designed and well-executed research methods, especially firsthand observation, interviews, and other qualitative techniques, all of which combined to produce a comprehensive picture of the actual situation concerning how the institution monitors and regulates research with biological HAZMAT. Third, the book analyzes internal policies, procedures, and practices, and reviews materials from other research institutions and federal guidelines for safeguarding biological materials. By doing so, the research is able to highlight several concerns over the safety and security for biological HAZMAT and offers sound recommendations based on these comparative analyses.

Dr. Valcik's research is thus timely, comprehensive, and important. This study makes a significant contribution to public policy by addressing an extremely crucial issue about which there can be little argument—the need for comprehensive guidelines, procedures, safeguards, and regulations concerning dangerous and hazardous biological research. This book will likely become mandatory reading for all university administrators charged with the responsibility to regulate scientific research.

Paul E. Tracy, Ph.D.
Professor of Criminal Justice
University of Massachusetts–Lowell
Lowell, Massachusetts

Preface

In 2005 I completed my dissertation on biological hazardous materials (HAZMAT) and research universities' compliance with the new Bioterrorism Preparedness and Response Act of 2002. The results from that research were both informative and alarming. The case study university was responsive to issues that were discovered and acted quickly to remedy many shortcomings that were reported to it. The dissertation was published in 2006 as *Regulating the Use of Biological Hazardous Materials in Universities: Complying with the New Federal Guidelines* (Mellen Press), which I thought originally would be a one-time publication. After the book was published I wrote a chapter regarding the new federal statute that required research centers that carried certain quantities of "chemicals of interest," or chemicals that could be used in criminal or terrorist activities, to report their inventories to the government. In 2012, I considered revising the book to incorporate the chapter I had written as well as examining what had occurred at the university in my study since 2005. Essentially, I was able to see if and how the HAZMAT situation had changed for the case study institution as well as for public organizations in general in terms of reacting to new federal statutes. Public organizations are usually slow to change in response to new situations and mandates due to inertia, culture, and bureaucracy itself. Conforming to the new realities of hazardous material regulation is no different from compliance with other new rules and regulations except that compliance with new HAZMAT guidelines could be very pricey for some public organizations. It is my hope that this second edition will assist administrators of research centers in dealing with federal compliance with HAZMAT and make research centers safer and more secure than ever before.

Since the Manhattan Project of the 1940s, there has been a substantial increase in research activities at American universities that involve radiological, biological, and chemical agents. Concurrently, higher education institutions and research centers have had to respond to new legislation and guidelines that were enacted to regulate the use of HAZMAT in research. The passage of the Patriot Act in 2001 and the Bioterrorism Preparedness and Response Act of 2002 has focused more attention on how research universities maintain compliance with the new federal guidelines. In 2004, the Department of Health and Human Services, Office of the Inspector General, audited the procedures and practices that govern the storage and use of select agents at several research universities. The audit found multiple violations at these universities, which indicated a widespread shortfall in compliance with current procedures for security and safety with selected agents.

This shortfall is due not to incompetence or malfeasance, but rather to the tension created between research faculty and university administrators caused by the effort to stay in compliance with governmental regulations. Faculty members consider the work they perform on their research projects their top priority as part of their function at an institution. Administrators have the responsibility to ensure that the institution is in compliance with federal, state, and local guidelines, while reducing liability and increasing safety for the organization. The tension between faculty and administrators often creates disjuncture between what ideally should occur and what is actually occurring with regard to compliance with governmental mandates.

The Patriot Act of 2001 and the Bioterrorism Preparedness and Response Act of 2002 were followed by the Homeland Security Chemical Facility Anti-Terrorism Standards of 2007. These three federal laws are in addition to the oversight faced by higher education institutions and research centers from the Environmental Protection Agency (EPA), which regulates hazardous waste; the Drug Enforcement Agency (DEA), which regulates controlled substances; and the Nuclear Regulatory Commission (NRC), which regulates atomic material. Although research activities have long been under federal oversight for radioactive material and hazardous waste, the new regulations provide oversight to biological and chemicals used in research activities.

This book is a case study of one university's policies and practices with regard to the procurement, use, storage, and disposal of biological, chemical, and radiological HAZMAT in the context of a changing internal structure and regulatory environment. The research utilized qualitative methods for gathering the data that allowed for analysis of the current situation with regard to how the institution conducted operations with biological, chemical, and radiological elements in the organization. Recommendations were formulated from gathering policies, procedures, and practices from other research institutions as well as utilizing federal guidelines for safeguarding HAZMAT used in research. The research highlights several areas of safety and security for HAZMAT that can be improved and makes recommendations based on those findings. This updated edition focuses on new developments with research centers and higher education institutions with regard to biological, chemical, and radioactive material, and industrial waste HAZMAT. This edition also includes a follow-up on the case study university, disclosing the university's progress in resolving the security and safety shortcomings discovered during the original research that was conducted between 2004 and 2005.

Acknowledgments

I would like to acknowledge the most important person in my life, my wife, Kristi, who has supported me throughout the process with love, patience, kindness, and encouragement. I would like to thank my dad, who has been and always will be with me in spirit. To my mother, I thank you for always being there for me, providing your love, a solid foundation, and faith in my abilities. To Erik and Heather, I am very grateful for all of the help you have provided to me while going through this process. To Jim, Chris, Amber, Amanda, and Jimmy, I appreciate all of the encouragement and caring you have given to me over the years.

I thank my co-workers who worked on development of the LTS© software: Danald Lee, Dr. Patricia Huesca-Dorantes, and Tarang Sethia. Twelve research assistants provided additional programming support for LTS throughout the years: Rajesh Ahuja, Mohit Nagrath, Dinikeshwari Byrappa Nagaraj, Ashwin Nayak, Rishi Raj, Shalu Agrawal, Roudra Bhowmick, Priyankar Datta, Ajeet Singh, Dhaval Shah, Seetha Kode, and Rasagna Ramireddy. Without their innovation and hard work none of this would be possible.

I thank my co-workers and supervisors for their good work during the initial research for this book as well as their continued assistance during the revisiting of the project. I thank Ted Benavides, who took time to review the manuscript and give feedback, which helped identify areas that needed to be worked on for this edition. I thank Craig Riggs for his insight and assistance in gathering private industry safety documents for this updated edition. I also thank my respondents for their time and effort during the interviewing process. I thank Jan Klein from the University of Wisconsin–Madison, Lee Zacarias from Georgia Tech, and Vincent Franconere from SUNY–Albany for responding to my inquiries of their university and departmental operations with HAZMAT. Last but not least, I thank Andrea Stigdon for all of her guidance, editing, and cover art provided on this book.

Author

Nicolas A. Valcik currently works as an associate director of Strategic Planning and Analysis for the University of Texas at Dallas and serves as a clinical assistant professor for public affairs for the University of Texas at Dallas. Nicolas received a doctorate degree in public affairs from the University of Texas at Dallas in 2005, a master's degree in the same field from the University of Texas at Dallas in 1994, and an associated degree in political science from Collin County Community College in 1994.

Prior to 1997, Nicolas worked for a number of municipalities, across different departments, as well as for Nortel. In 2006, Nicolas authored *Regulating the Use of Biological and Hazardous Materials in Universities: Complying with the New Federal Guidelines*, published by Mellen Press. He has served as editor of three volumes of *New Direction for Institutional Research* (Volumes 135, 140, and 146), in addition to writing numerous articles and book chapters on institutional research topics and homeland security issues. Nicolas specializes in several areas as both a researcher and a practitioner: higher education, information technology, human resources, homeland security, organizational behavior and emergency management.

1

Regulatory Change and Organizational Responses: A Case Study of the Procurement, Use, Storage, and Disposal of Hazardous Materials (HAZMAT) in a University Environment

INTRODUCTION

The purpose of this research is to examine one research institute's policies and practices with regard to the procurement, use, storage, and disposal of biological, chemical, radioactive and industrial wastes that are hazardous materials (HAZMAT) in the context of a changing internal organizational structure and a changing regulatory environment. In addition, the research reviews HAZMAT policies at other selected universities, examines issues common to research organizations, and recommends policies and practices that might improve HAZMAT security and safety. Biological HAZMAT includes select agents, controlled substances that induce biological ailments (e.g., Parkinson's disease), animals, plants, blood-borne pathogens, human waste by-products, tissue samples, and other waste produced from biological research experimentation. Chemical HAZMAT consists of materials used in research that can be toxic, are controlled substances, corrosive, poisonous, or potentially explosive (e.g., cyanide, sulfuric acid, and asbestos). Radiological HAZMAT includes materials that use radioactive isotopes for research purposes (e.g., U-235 and PU-39). HAZMAT waste is defined as by-products from biotoxins, chemicals, or radioactive isotopes that have been discarded after being used in research activities and can present a danger to living organisms.

Although research universities and research centers may have unique histories and workflow processes with regard to HAZMAT, it appears that environmental pressures eventually force research institutions to standardize operations to meet regulations. Change will occur at different rates for each individual institution due to each university having its own specific research operations and unique faculty characteristics contending with HAZMAT. The types of HAZMAT used for research will determine the risk factors for noncompliance with government regulations.

To ensure compliance, many institutions will need to change the structure and processes of the organization to provide adequate support for researchers using HAZMAT. Fundamentally, universities need to protect critical physical assets, information, and materials while also ensuring that researchers can have the freedom they need to engage in creative endeavors. At their normative extremes, the excesses in security could prove to be overly bureaucratic and stifling, whereas at the other extreme, individual laboratory controls can break down, resulting in materials misplaced and experiments altered. Two factors influence current and future policy. First, the scientific ethos and method provide a powerful directive for the researcher to desire control over his or her laboratory. This plus local policy was the predominate means by which science advanced until the later 1800s.

Increasingly, the federal government, in response to perceived problems, threats, or crisis, has passed legislation that allowed federal agencies to create guidelines that impact research universities' ability to perform research operations. The rise of federally sponsored research programs began with the National Institutes of Health (NIH) in 1930. Over time, more federal agencies became involved in research programs related to health, safety, and security for biological elements. The Drug Enforcement Agency (DEA), which was established in 1973, extended its jurisdiction over controlled substances used in research. The development of new technologies spurred more regulations and more oversight by federal agencies. The Atomic Energy Commission (AEC) is an example of a new agency created specifically to oversee the proper use and handling of radioactive material.

With federally sponsored research programs came federal regulations, first with federal guidelines from the NIH and later from a multitude of federal agencies. Universities began facing an increasingly complex and changing regulatory environment. At the same time, more sensitive and potentially hazardous materials made their way into research projects. Even so, behavior patterns for many universities did not embrace changes in policies, procedures, and practices that are perceived as a possible infringement to research activity—particularly those with small research efforts and those lacking already established administrative structures. This created two main concerns for universities: (1) new regulations that rendered existing science facilities obsolete and (2) new laboratory procedures that caused personnel trained in older methods and practices to become noncompliant.

The public, too, became more aware of the dangers associated with research done at universities:

> In Boston, activists are trying to halt construction of a $168 million lab at Boston University, to be built in the city's south end. They fear something akin to what health inspectors suspect occurred recently in China: that SARS escaped a Beijing laboratory and made its way into the Chinese heartland, contributing to the latest eruption of the sometimes fatal disease. (Elias 2004)

The two maps in Figure 1.1 and Figure 1.2 illustrate the major incidents that have occurred with HAZMAT at research centers and research universities since 1951. These two maps show the number of incidents that have occurred with all types of HAZMAT used at public research centers and research universities since 1951. In addition, the geographic information system (GIS) maps illustrate how many incidents have occurred in the northeastern part of the country, where dense population centers are located, compared to midwest, southern, or western regions of the country.

Thus, a multitude of factors—the increasingly federalized research agenda and the regulatory strictures attached, the presence of research universities in urban areas, the increasing perceived virulence of research materials, the rise of the new animal antivivisectionists and their fellow travelers as well as sporadic but effective public outcries—pressed for change and dictated the need for striking a balance between HAZMAT security sufficient enough to assuage public concerns and protect assets while simultaneously encouraging independent thought, experimentation, and the sharing of results. Incidents that are not reported properly or are perceived to be "covered up" can create fear and distrust of institutions that use HAZMAT for research. Texas A&M University in 2006 failed to report in a timely manner the infections of three workers who were in contact with Q fever (a select agent) to the Centers for Disease Control and Prevention (CDC) as required by federal statute (Ramshaw 2007). In

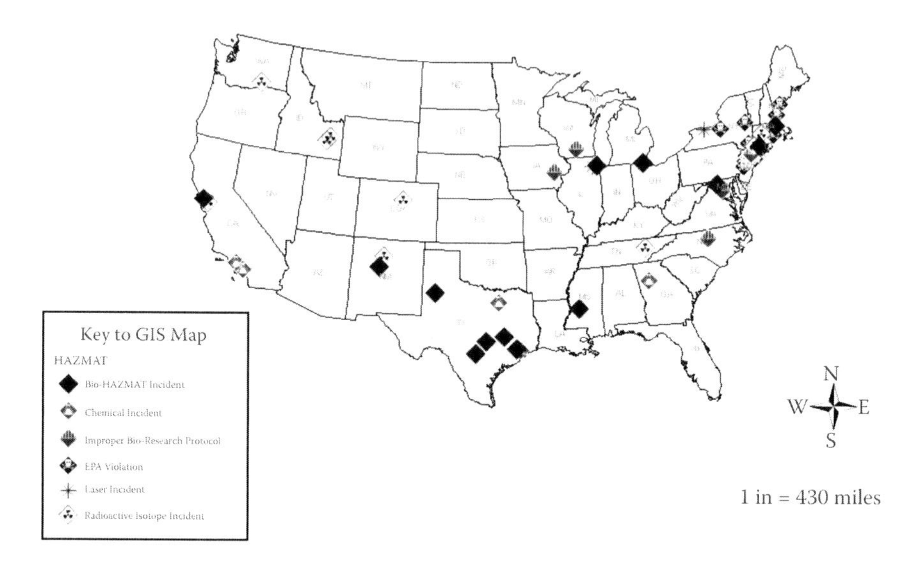

FIGURE 1.1
1951–2009: HAZMAT incidents in United States higher education and government research centers.

that same year, Texas A&M had another worker infected by Brucella while the worker was cleaning a chamber (Ramshaw 2007). In both cases Texas A&M did not report the incidents to the CDC for at least a year after the incidents had occurred (Ramshaw 2007). Thus even a prestigious research institution like Texas A&M can endanger its reputation if such incidents are not reported and managed properly.

Rhodes (1986), in his treatment of the development of the atomic bomb, typifies another dimension of the interactions between the need for security and the scientist's need for interaction, especially when dealing with material on the edges of scientific knowledge. General Leslie R. Groves focused his efforts on absolute security for the Los Alamos Research Laboratory. J. Robert Oppenheimer, the principal investigator behind the nuclear research, maintained his focus on scientific progress. The disparate approaches taken by Groves and Oppenheimer emerged from opposing ideologies. From Grove's perspective, the free exchange of information among colleagues that was the norm in other institutions could not be permitted at Los Alamos due to the high security risk from the country's many enemies. The "Los Alamos model" is an archetypical example of the social control and regulatory issues involved in research on hazardous materials.

FIGURE 1.2
1951–2009: Detailed map of HAZMAT incidents in northeastern United States higher education and government research centers.

The Cold War hastened the development of more powerful and deadly weapons such as the hydrogen bomb, the intercontinental ballistic missile, tactical nuclear weapons, and chemical-biological weapons that demanded a much higher degree of protection for national security reasons. As the government awarded more funds for weapons research, researchers expanded their activities into other forms of scientific endeavors. The search for alternatives to weapons of mass destruction spawned research into (then legal) psychedelic drugs, ultimately aiding in the creation of the "hippies" and a host of new federal regulations aimed at laboratory use and documentation of controlled substances. As stated by Howard S. Becker:

> In the early 1960's, Timothy Leary, Richard Alpert and others began using it with normal subjects as a means of "conscious expansion." Their work received a great deal of publicity, particularly after a dispute with Harvard authorities over its potential danger. (Becker et al. 1968, 273)

Research advancements into different fields of study increase the risks for accidents, criminal acts, or a potential breach of national security. The types of hazardous materials stored and used at universities and colleges are under new scrutiny. Before, a chemistry laboratory might only have basic substances such as sulfur, iodine, and magnesium. Today, a well-equipped research laboratory could stock anything from morphine and other controlled substances to hydrochloric acid and other extremely corrosive materials.

When higher education institutions receive substantial research money, the granting agency can demand greater involvement in research activities and better protection of assets. Some institutions are not prepared to accept such invasiveness. For example, when the National Security Agency (NSA) offered the Massachusetts Institute of Technology (MIT) a $400,000 grant on the condition that foreign students could not work on the research project, MIT rejected the contract (Twohey 2003).

There are many other examples that point to a shift in the locus of control over research, especially research involving controlled and potentially dangerous materials. Consequently, internal control structures of universities are under pressure to change and be in alignment with federal (or sometimes private) regulatory strictures that are attached to research funds (see Olswang and Lee 1984). The passage of the Patriot Act of 2001 and the Bioterrorism Preparedness and Response Act of 2002 ushered in a new era of federal oversight that increased the number of guidelines over research universities. As of 2004, the Department of Health and Human Services, Office of the Inspector General had started to audit universities on their security and safety measures over the use and storage of select agents. The inspectors discovered several shortfalls with regard to policies, procedures, and practices for handling select agents. The Environmental Protection Agency (EPA) has also performed audits on a number of universities and colleges on the proper handling and disposal of waste. This agency also discovered a number of shortfalls at several institutions for improper waste disposal. The discovery of widespread inadequate biological HAZMAT controls leads to the issue of how research universities are structured for controls of research that is in progress.

New chemical HAZMAT regulations include the Homeland Security Chemical Facility Anti-Terrorism Standards 2007, which requires research centers and research universities to report certain chemicals to the federal government if their inventory of those chemicals exceeds a specific threshold limit. The new inventory requirements, however, can be a double-edged sword. This new regulation requires organizations to compile an inventory list and transmit that list to an external agency. Although maintaining a chemical inventory is ideal for inventory control and as a tool to aid first responders, the disadvantage is that the list could potentially fall into the wrong hands, which would give anyone with criminal intent the ability to selectively target certain facilities or specific types of chemicals.

The following case study of one research institution (described as "the University") will demonstrate how the institution's unique history, its environment, and its place in the evolving nature of federal research led to the existing conditions of its HAZMAT. The University, originally a research think tank, evolved into a public research university with an enrollment of thousands while simultaneously experiencing a substantial increase in the quantity and complexity of its research. The University's environment transformed from an isolated, rural area into an expanding suburban city. Federal regulation of materials exerted greater influence over the University's activities as the surrounding area filled with homes and private businesses, placing citizens much closer to facilities where hazardous materials were stored and used. The degradation of adequate public funding for the University's growing enrollment and research needs was mirrored by the degradation of the University's primary science building as more laboratories were added without the necessary upgrades to comply with new federal regulations.

The organizational culture also underwent change as instructional responsibilities were added to the faculty's research duties. Adequate security became difficult to maintain as increased foot traffic due to the construction of new classrooms carved from old laboratory spaces resulted in students walking through corridors once restricted only to researchers. The investigation into the evolution of the University's HAZMAT condition utilizes several organizational theories. By utilizing the research approach known as grounded theory, existing HAZMAT conditions can be analyzed to determine why and how these conditions evolved within the University. Life cycle theory indicates how organizational drift could occur as the amount of research performed at the institution outpaced available monetary resources, and why the loci of control shifted away from investigator-controlled research centered. When the University expanded in size and research capabilities, more federal regulations forced the organization to change existing policies and procedures to remain compliant with federal guidelines.

The research addresses how organizational drift led to the misalignment of the University's policies, practices, and procedures away from current federal regulations due to the University's expansion. The University was unable to accommodate many of the recent federal guidelines because existing resources were applied toward developing its research capacity while new resources were funneled into expanding classroom instruction. Eventually this shortcoming would cause a shift from investigator-controlled research to a more organization-centered research model. Finally, agency theory is relevant for explaining another element of how organizational drift and decoupling caused misalignment with current federal regulations.

Because the University was originally a research think tank, a logic of confidence in the researchers existed in the culture of the organization, which did not provide oversight to the researcher's activities. Administrators and faculty assumed that researchers had enough knowledge about their work to operate safely in research laboratories. As research expansion at the University outstripped existing resources, the researchers were left to work on their projects with little or no resources to support their existing operation much less accommodate new federal requirements on an existing infrastructure that was not going to be upgraded.

Qualitative methods such as conducting security surveys and personnel interviews, researching best practices, and analyzing archived HAZMAT documentation were used to gather data throughout the duration of this research. To construct a robust timeline that adequately explained the existing HAZMAT situation for the University, it was necessary to use these different sources to triangulate the data. Best practices were collected through documentation of other universities' policies and procedures, guidelines from federal agencies, and interviews of personnel employed at other universities. By analyzing this documentation, it can be determined if the University's current operations meet the minimum standards of other entities. The participant observer research method was utilized to gain insight on operational aspects for HAZMAT. By working with administrative units that exercised jurisdiction over HAZMAT policies and procedures at the University, additional data was obtained that would have been difficult to obtain independently. Furthermore, the participant observer method enabled the implementation of some of the recommendations proposed in this study that will improve safety and security at the university.

Data collected from these research methods were critical in formulating a grounded theory from which the topic could be analyzed. Initial data gathered through unobtrusive observations enabled the formulation of survey questions that would determine how and why the current HAZMAT conditions existed. It was necessary to determine if the facilities that contained HAZMAT had always operated according to certain methodologies or did the existing situation evolve from one that once was more stable and secure. (The data gathered in the security survey would reveal that the situation had, in fact, evolved due not only to a research-centered institution adding instruction to its mandate but also to the inclusion of research programs housed in facilities that were not originally constructed to contain them.)

Finally, recommendations are proposed to improve safety while using and storing biotoxins, chemicals, radioactive material, and industrial waste, and to improve overall security at the University. Some of these recommendations are easier to implement than others and are dependent upon which priorities the University wishes to emphasize. The research does provide the University a number of issues that could be improved to make the environment a safer and more secure location to perform biological research activities. Other universities, such as Georgia Tech, are currently attempting to improve biohazard safety and security by using new tools and techniques that might also prove invaluable to the University. By implementing key improvements in safety and security, the

University might also find that it can more easily obtain research grant money and satisfy both state and federal safety requirements. To understand compliance issues, it is necessary to review the history and impact of selected federal agencies responsible for oversight of research universities' research activities.

2

Rise of Federal Agencies' Influence over Research Institutions

BACKGROUND OF FEDERAL GOVERNMENT INVOLVEMENT WITH RESEARCH UNIVERSITIES

To understand why the federal government is so concerned about the regulation of hazardous materials (HAZMAT), it is important to accurately define what constitutes HAZMAT. Southern California Edison, one of the largest electric utilities in the United States and the largest subsidiary of Edison International, defines HAZMAT as both a material and as a by-product:

> Hazardous Material: Any material that because of its quantity, concentration, or physical or chemical characteristics, poses a significant present or potential hazard to human health and safety or to the environment if released into the workplace or the environment. Hazardous materials include, but are not limited to, hazardous substances, hazardous waste, and any material which a handler or the administering regulatory agency has a reasonable basis for believing would be injurious to the health and safety of persons or harmful to the environment if released into the workplace or the environment (California Health and Safety Code, Section 25501 [o]). A number of properties may cause a substance to be considered hazardous, including toxicity, ignitibility, corrosivity, or reactivity.
>
> Hazardous Waste: A waste or combination of waste which because of its quantity, concentration, or physical, chemical, or infection characteristics, may cause or significantly contribute to an increase in mortality or an increase in serious irreversible or incapacitation-reversible illness; or pose a substantial present or potential hazard to human health or the environment, due to factors including, but not limited to, carcinogenicity, acute toxicity, chronic toxicity, bioaccumulative

> properties, or persistence in the environment, when improperly treated, stored, transported, or disposed of or otherwise managed (California Health and Safety Code, Section 25141). California waste identification and classification regulations are found in Title 22 of the California Code of Regulations. (Southern California Edison 2008)

The four main categories of HAZMAT are chemical, biological, radiation, and waste. Chemical HAZMAT is now regulated under the new guidelines set by the Department of Homeland Security (DHS). Chemicals that are considered controlled substances also fall under the jurisdiction of the Drug Enforcement Agency (DEA). Furthermore, waste produced from research and operational activities must be disposed of according to Environmental Protection Agency (EPA) regulations.

With the advent of nuclear power in the 1940s, the federal government formulated guidelines to control materials and research activities for safety and security. A myriad of federal agencies, including the Department of Defense (DOD), National Aeronautics and Space Administration (NASA), and the Department of Agriculture, set guidelines for researchers using (or producing) HAZMAT during research experiments. The guidelines set forth by the federal agencies do not include regulations that can also be imposed by state or local agencies. Recent terrorist attacks, lost biospecimens, and the arrest of international students as terror suspects are responsible for the creation of a new federal agency—the Department of Homeland Security—to protect the infrastructure and assets of the United States (U.S. Congress 2002). The Department of Homeland Security, the National Institute of Justice (NIJ), and the Federal Emergency Management Agency (FEMA) have all recently issued requests for funding proposals (RFPs) to protect public infrastructure and personnel from criminal and terrorist actions. A NIJ RFP states:

> Protective Systems Technologies. The goal of this program is to provide tools that help protect individuals and locations. This could include concealed weapon detection systems, through-wall sensors, tracking and monitoring systems for individuals, and protective equipment for personnel. (National Institute of Justice 2003)

NIJ views the protection of public personnel, assets, facilities, and capital from an act of criminal intent or terrorism as a widespread point of vulnerability. A FEMA RFP, through the Department of Homeland Security, stresses the importance of the well-being of public universities' assets and infrastructure. In its RFP, FEMA (2003) states:

> FEMA will provide Pre-Disaster Mitigation (PDM) funds to assist universities, through the state and local governments, to implement a sustained pre-disaster natural hazard mitigation program to reduce overall risk to facilities, research assets, students and faculty.

FEMA clearly states the intentions of the grant proposal. As the Department of Homeland Security (2003) notes in an RFP issued by the department:

> 5.3 Infrastructure Protection (IP) Mission Area/Subgroup—The Infrastructure Protection (IP) Subgroup identifies and pursues user requirements for the protection and assurance of critical government, public, and private infrastructure systems required to maintain the national and economic security of the United States.

As can be seen by the Department of Homeland Security RFP, finding a method to protect infrastructure is important enough to warrant its own category of funding. Congress is also considering a bill that will force public and private entities to tighten security (Hanson 2004). Beginning in the health regulatory sector and ballooning to other agencies, there has been a proliferation of regulations that require university compliance.

In 2001 the Patriot Act was passed in response to the terrorist attacks of 9/11. The Patriot Act requires that individuals who are considered "restricted" be denied access to select agents as listed (Youngers and Norris 2004). The Patriot Act has been amended for security of biological agents under the Public Health Security and Bioterrorism Preparedness and Response Act of 2002 (U.S. Congress 2002). The Patriot Act amendments direct the United States Department of Agriculture (USDA) and Department of Health and Human Services (HHS) to compile regulations for institutions that use, store, or transport any type of biological agent or toxin that is deemed "a threat to animal or plant health and to animal or plant products" (U.S. Congress 2002). According to the amendment, the U.S. Attorney General is required to perform a background check on anyone who will have access to such elements (U.S. Congress 2002). The Centers for Disease Control and Prevention (CDC) and the Animal and Plant Health Inspection Service (APHIS) have a list of select agents, pathogens, and toxins that are considered regulated by the Patriot Act amendment (Youngers and Norris 2004).

In 2007 Congress passed the Homeland Security Chemical Facility Anti-Terrorism Standards, which targeted how research centers and research universities accounted for specific chemicals that were over a certain threshold amount. This statute mandates that the Department of Homeland Security (DHS) create guidelines that will require research centers and research universities to report to a government agency those chemicals that could be used in criminal or terrorist activities. Before this statute was passed there was no such reporting requirement for those types of chemicals specified by DHS. Unless the research center or research university kept an active inventory of chemicals, there was no accurate way for auditors or first responders to know where the chemicals were located, the quantities of chemicals that were stored at a certain location, what the chemicals were being used for, the rate of consumption for those chemicals, and ultimately who was responsible for those types of chemicals. The lack of information can pose problems for inventory control or emergencies that may occur in a particular facility.

HISTORY OF FEDERAL AGENCIES

The Drug Enforcement Agency (DEA) has a long history dating back to 1915 of operating under the Department of Treasury as a part of the Bureau of Internal Revenue (DEA 2004). As seen in Figure 2.1, the Bureau of Narcotics and Dangerous Drugs operated under the Department of Justice in 1968. In 1973, the DEA that is currently operating was established from the Bureau of Narcotics and Dangerous Drugs (DEA 2004).

The Atomic Energy Commission (AEC) was established with the passage of the Atomic Energy Act of 1946. With the advent of commercial nuclear energy, and the perceived inadequacy of controls and safety in AEC's programs, the Energy Reorganization Act of 1974 created the Nuclear Regulatory Commission (NRC) to protect public health and safety (Figure 2.2). As stated by the NRC (2004):

> One such issue was the protection of nuclear materials from theft or diversion. This became a prominent question after the 1970s in response

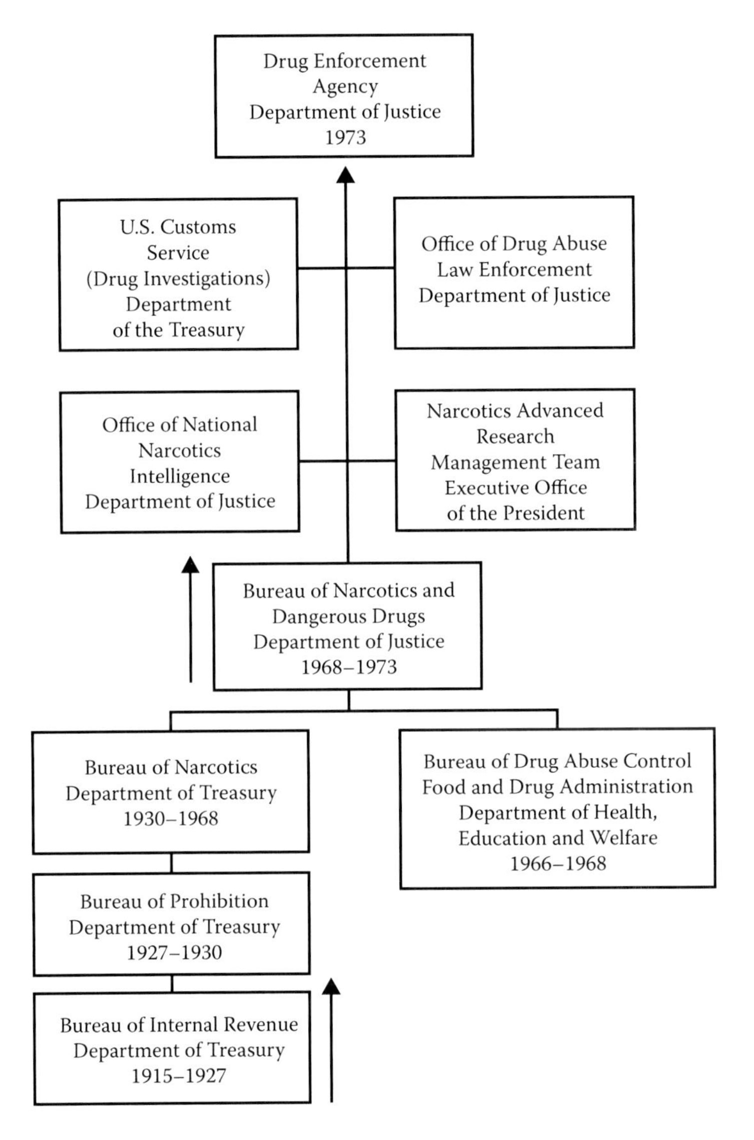

FIGURE 2.1
History of the Drug Enforcement Agency. (From Drug Enforcement Agency, 2004, "DEA genealogy," http://www.usdoj.gov/dea/history.htm.)

to growing concern that nuclear materials could be obtained by terrorists or nations seeking to build atomic weapons. The NRC also devoted a great deal of attention to the safety of depositories for the disposition of high-level and low-level radioactive waste, which was a matter of public fear and bitter political controversy.

Atomic Energy Act 1946 establishes the AEC

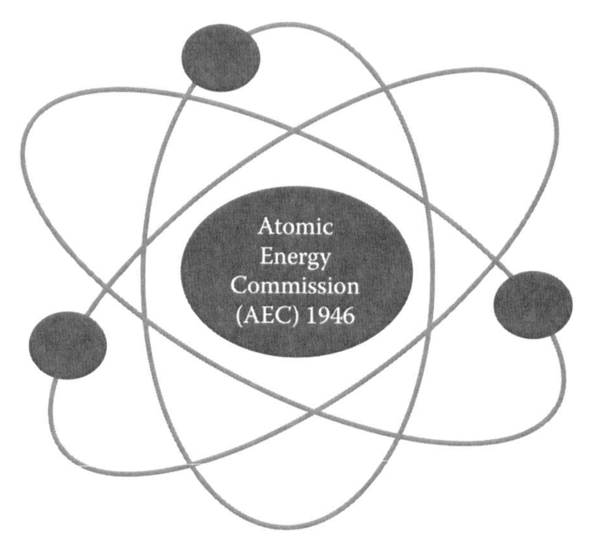

Atomic Energy Act 1954 allows for Commercial Nuclear Power

The Energy Reorganization Act of 1974 establishes the NRC

FIGURE 2.2
History of the Nuclear Regulatory Commission.

The Environmental Protection Agency (EPA) was established in 1970 to promote clean air and water in the environment (EPA 2004b). The EPA was composed of the following functions from different agencies:

- Federal Water Quality Administration—Department of the Interior
- Pesticide Studies—Department of the Interior
- Bureau of Solid Waste Management and Bureau of Water Hygiene
- Some functions from the Bureau of Radiological Health of the Environmental Control Administration—Department of Health, Education and Welfare
- Some functions from the Food and Drug Administration—Department of Health, Education and Welfare
- Ecological Studies from the Council on Environmental Quality
- Some functions from the Atomic Energy Commission and Federal Radiation Council
- Functions from the Agricultural Research Service—Department of Agriculture

As seen from the functions list, EPA has a broad and sweeping mandate that incorporated many areas from several different agencies.

The National Institutes of Health (NIH) was initially created as the Marine Hospital Service (MHS) in 1798 to service merchant seamen. In 1902, the Biologics Control Act was passed by Congress to regulate vaccines and antitoxins that reorganized the MHS into the Public Health and Marine Hospital Service (PH-MHS). In 1912, the PH-MHS became the Public Health Service (PHS) and began to research diseases that were related to the pollution of water supplies. With the passage of the Ransdell Act in 1930, the organization became known as the National Institute of Health and was tasked with promoting public funding for medical research. In 1944, the National Institute of Health was given authorization to begin clinical research. In 1948, with the onset of increasing problems with heart disease, the National Institute of Health became the National Institutes of Health. As stated by NIH: "Since 1985, NIH officials have conducted unannounced site visits to spot-check animal facilities and laboratory research programs" (NIH 2004).

The Department of Homeland Security (DHS) was created in 2002 with the 2002 Homeland Security Act in response to a lack of perceived unity in both intelligence and communication among the various law enforcement agencies. The DHS is currently composed of twenty-two federal departments with the responsibility to protect the United States from terrorist and criminal activity (DHS 2011).

MAJOR FEDERAL GUIDELINES APPLICABLE TO RESEARCH CENTERS AND RESEARCH UNIVERSITIES

Given the research that is performed at research centers, universities, and colleges, a number of guidelines have been instituted by a variety of federal agencies (see Appendix A). The DEA, for instance, has guidelines regulating controlled substances in use for experiments on animals and human subjects. The NRC is concerned with radioactive research and the control of radioactive material. NIH has guidelines that regulate biological specimens and biohazards with regard to research activities. NIH and CDC also have guidelines on how research laboratories must be constructed dependent upon the level of hazardous research being conducted at the facilities. DHS has guidelines concerning particular chemicals that are over a certain threshold and the EPA is focused on regulating waste produced by research centers and research universities.

With the passage of the USA Patriot Act os 2001 and the Bioterrorism Preparedness and Response Act of 2002, revised and updated guidelines (posted March 18, 2005) are now in place with the CDC and the USDA with regard to biological microbes and toxins. Security guidelines for biological research have not been posted for universities from the CDC and the USDA but are forthcoming (Brainard 2005). The new regulations were devised in response to the anthrax attacks that occurred in 2001 (Brainard 2005).

In a March 2004 report titled "Summary Report on Select Agent Security at Universities" from the Department of Health and Human Services, Office of Inspector General, eleven universities were audited for security procedures of selected agents used or stored at those institutions (Department of Health and Human Services 2004). All eleven universities were failed in the audit for shortcomings in physical security and inadequate inventory control procedures. Five of the eleven universities had shortcomings with information technology security and data integrity. At least half of the institutions failed to have procedures to identify and bar restricted persons (defined by the USA Patriot Act of 2001) from selected agents (Department of Health and Human Services 2004). The auditors used generally accepted government auditing standards to perform the security audit. The physical security weaknesses noted in the report are as follows:

- Uncontrolled issuance of hot lab door keys
- Unauthorized entry into "hot labs"
- Once intruders accessed the building, intruders had unobstructed access to the floors with hot labs
- Unauthorized removal of selected agents
- Storage of hot lab keys in open drawers
- Unlocked security doors
- Use of breakable glass in security doors
- Unsupervised, unimpeded access to freezers located in open areas
- Unlocked freezers in hot labs
- A lack of closed-circuit television cameras
- Nonenforcement of the use of identification badges

The auditors could not determine whether proper procedures had been used for compliance of the Select Agent Transfer Regulation due to inadequate inventory control (Department of Health and Human Services 2004). As stated in the report:

> We noted, however, that principal investigators and their assistants generally did not maintain select agent records, on the date that vials were placed into or removed from inventory, the source and destination of shipments, or the total inventory on hand. We also noted that the universities did not require principal investigators to physically count select agent inventories to verify their accuracy (Department of Health and Human Services 2004).

Due to the lapses in security and inventory control, the report also pointed to the universities' lack of knowledge with compliance to security for select agents. As stated by the auditors:

> Some universities cited concerns about the lack of criteria in the security area and the funding required to implement the necessary improvements. Some of the universities said that the BMBL [*Biosafety in Microbiological and Biomedical Laboratories*] was only a general guideline and not a requirement—a comment to which we took exception. Although issued as a guideline, the BMBL had been formally incorporated into the select agent regulation (42 CFR § 72.6(a)(5)) and was in effect at the time of our reviews (Department of Health and Human Services 2004).

The second round of compliance audits on fifteen universities with select agent regulations was performed by the Department of Health and Human Services, Office of the Inspector General, between November 2003 and November 2004 (Department of Health and Human Services 2006). Out of the fifteen universities only four universities passed criteria set forth by the auditors as having proper controls in place for complying with federal guidelines (Department of Health and Human Services 2006). The other eleven universities failed the auditors' criteria in the following aspects:

- Eight institutions had shortcomings in inventory control and incomplete or inaccurate information for users who accessed that information.
- Six institutions had shortcomings in processes and procedures to control electronic access keys to areas that contained select agents.
- Six institutions had shortcomings in their security plans and/or incomplete risk analysis in regard to select agents.
- Three institutions had not properly documented training procedures for workers with select agents or did not provide workers with proper training to work with select agents.
- Three institutions had inadequate emergency response plans.

As stated in the Department of Health and Human Services (2006) report: "However, one institution disagreed with our finding that it lacked accountability for select agents, even though the university had not verified its inventory records in 8 years."

Both audit reports illustrate the shortcomings that institutions face in complying with federal statutes on select agent safety and security. Since 2001, the reports also show that universities and colleges still lag behind in federal compliance issues in regard to select agent statutes six years after 9/11 occurred.

In addition to the CDC, NIH, and USDA regulations, research centers and universities must also adhere to EPA regulations for HAZMAT disposal. The EPA is concerned with HAZMAT disposal and usage at facilities to protect the health of students, faculty, staff, employees, and the environment. All of the federal regulations have been gradually tightened since the early 1980s. Research facilities that were constructed before the passage of such guidelines either had to be demolished and a new facility constructed to accommodate the research activities or had to be modified to be compliant with the federal guidelines. Other federal agencies have more stringent regulations based on contracts or grants that the college or university may be actively researching (e.g., a Department of Defense contract). EPA in particular released information in 2000 that several universities had either been fined for violations or been found to be in noncompliance with EPA guidelines.

> The EPA Regions have found specific examples of noncompliance at universities and colleges that include improperly handling and disposing of hazardous waste materials, boilers and furnaces that do not meet clean air regulations, inadequate monitoring of underground storage tanks, sewage treatment facilities that are not operating properly, and proper abatement of lead-based paint and asbestos. … Colleges and universities, as well as military educational institutions such as the Air Force Academy and West Point, are required to comply with all applicable environmental requirements like their counterparts in the regulated community to create a safe haven for human health and the environment. Violating the environmental requirements can be costly, for example the University of Hawaii recently paid $1.8 million in civil penalties for violating federal law by poorly managing laboratory waste. (Howell and Hanif 2004)

The EPA has been increasing the number of on-site audits to universities in the last few years. In 1997, Boston University and Brown University were both fined hundreds of thousands of dollars for EPA violations. As stated by Rene Henry, a governmental relations director for the EPA, "Our inspectors have not been on one campus where they have not found serious problems" (Associated Press 2001). The EPA has fined a number of colleges and universities across the country due to violations of federal guidelines (Associated Press 2001).

In addition to federal guidelines, universities must also face state regulations and agencies. Such was the case at Texas A&M University when an apartment blast killed two people and injured two others. The blast was caused by a leak in the natural gas line. This event initiated an inspection by the state's Fire Marshal Office (Associated Press 2004a). This incident further points to increased regulations and inspections that are being placed on universities and colleges. As stated by the Fire Marshall's Office (Texas Department of Insurance 2004):

> The Texas Department of Insurance, State Fire Marshal's Office (SFMO), is required to periodically report on the progress of institutions of higher education in remedying the fire safety issues identified by the State Fire

> Marshal's Office. (House Bill 1, Article III, Section 39. Fire Safety Projects at Institutions of Higher Education)

Recently the University of Texas at Austin was accused of ignoring shortfalls in safety and federal compliance guidelines by a former employee who was a veteran national security expert (Associated Press 2008). In 2008 a former lab safety expert stated he was terminated from his position because he informed administrators that safety standards were not being followed and that the university was disregarding federal standards for reporting research activities in regard to laboratory accidents over a seven year period (Associated Press 2008). When incidents of this nature are not reported by any research organization, it should be cause for concern that there may potentially be other shortfalls that exist within the organization's policies and procedures that could create safety and security concerns.

The federal government maintains an interest in how research centers and universities manage chemical HAZMAT because chemicals used in traditional research laboratories can be stolen and used to commit crimes or acts of terrorism. Research centers and higher education institutions that use controlled substances like morphine and LSD in animal research must contend with the DEA, which maintains strict regulation on these types of controlled substances. New statutes from Homeland Security categorize certain chemicals as potential danger for release, theft, or diversion or sabotage or contamination, and these chemicals are denoted as "chemicals of interest" if the amount of the chemicals is above a threshold limit at a facility (Department of Homeland Security 2007). According to DHS guidelines, the following chemicals must be reported: chlorine (1,875 lbs threshold limit), chloroform (15,000 lbs threshold limit), mercury fulminate (2,000 lbs threshold limit), and nitrogen trioxide (any amount threshold limit). These chemicals are only a sample of what can be found at higher education institutions that are also listed on the DHS chemicals of interest appendix.

In the past, data on HAZMAT used and stored at research centers or universities might have been recorded on paper, if at all. Given the preponderance of HAZMAT on most university property and the increasing complexity of federal reporting mandates, the decision to record this information electronically seems to be the more practical. An electronic database can be accessed more readily by first responders in times of crises, can be restricted by information security personnel so that only individuals with proper clearance can access the data, and can enable environmental health personnel to track HAZMAT across the university to exact locations or individuals. However, once HAZMAT information is electronically recorded, information security and risk assessment take on much greater importance. As stated by Texas Administrative Code, Title 1—Administration, Part 10—Department of Information Resources, Chapter 202 Information Security Standards, Subsection C—Security Standards for Institutions of Higher Education, Rule §202.72, Managing Security Risks (State of Texas 2008):

> (1) High risk annual assessment—Information resources that;
> (A) Involve large dollar amounts or significantly important transactions, such that business or government processes would be hindered or an impact on public health or safety would occur if the transactions were not processed timely and accurately, or
> (B) Contain confidential or sensitive data such that unauthorized disclosure would cause real damage to the parties involved, or
> (C) Impact a large number of people or interconnected systems.

According to this Texas statute, all information resources, inclusive of electronically recorded chemical HAZMAT data, must be secured and treated as high-risk information. Chemical HAZMAT data can impact transactions and processes, inflict damage, and impact large numbers of people or interconnected systems.

HAZMAT, including chemical HAZMAT, is also considered an asset to a public organization. For example, ammonium nitrate, a chemical widely used as a fertilizer, has monetary value, which requires that it be inventoried for cost considerations. Ammonium nitrate is also a dangerous chemical that, when improperly used, can become explosive, causing real harm to parties and disrupting key systems (e.g., Oklahoma City Alfred P. Murrah Federal Building bombing in 1995). Hence chemical HAZMAT itself, not just HAZMAT data, falls under Texas statute.

RISK ASSESSMENT: HAZMAT AT RESEARCH CENTERS AND HIGHER EDUCATION INSTITUTIONS

The public health of the community must be considered when assessing the risks involved with HAZMAT safety and security. These considerations provide ample reason to track HAZMAT to their locations and to secure HAZMAT properly. If a fire or a chemical disbursement occurs in a facility known or suspected to contain chemical HAZMAT, firefighters and other first responders will need to be informed of what those chemicals are to protect themselves and others, and to properly contain the emergency situation. The same scenario can also be true for waste, radioactive isotopes, or biotoxins that could potentially be in a facility, which could injure or kill first responders entering that facility. A detailed and updated HAZMAT inventory is necessary to ensure that first responders can correctly assess the threat potential of the situation, conduct proper evacuation procedures if necessary, employ adequate protective gear, and select the best methods for containing the emergency.

An emergency situation involving chemical HAZMAT, for example, should not be taken lightly. A typical research institution often uses and stores highly dangerous chemicals, such as chlorine gas, that have the potential to kill or injure people on a smaller scale. An extreme example of the potential destructive power of a large chemical release is the 1984 incident in Bhopal, India, where chemical exposure from a Union Carbide Plant killed approximately 3,800 people and injured several thousands more (Union Carbide 2007). A smaller-scale event still retains the potential for death and injury to the campus community as well as to residential and business areas that surround most universities. The case study university, for example, is surrounded by a residential area that includes small businesses and public schools, which might necessitate a significant public evacuation in the event a large-scale, airborne dispersal of dangerous toxins.

One specific challenge facing higher education institutions is the need to secure facilities while maintaining an open campus. Unlike national research laboratories such as Los Alamos, higher education institutions have other missions besides research that they must accomplish, such as instruction, public outreach programs, and medical services. Even at the federal level, security can be an issue, as was recently discovered at the CDC biological laboratory in Atlanta, which was cited for lax security in restricted areas of the facility. The size of a typical college or university campus (or multiple campus locations) coupled with the fact that research and medical services are frequently in multiple facilities that have mixed uses can make securing HAZMAT extremely difficult. If an emergency does occur (e.g., fire), safety is also a bigger concern than if the HAZMAT was concentrated in just one or two buildings. In an effort to better secure their facilities, a number of campuses have begun to concentrate their research facilities in separate buildings. Massachusetts Institute of Technology (MIT), for example, has recently been relocating classified research away from its main campus to improve security (Campbell 2002). In comparison, a private corporation, such as Raytheon, with a high level of classified research, has a comprehensive safety, security, and inventory program to control research activities (Raytheon 2012). Raytheon's policies and procedures, for example, require four aspects for research activities: design the research for minimum risk, incorporation of safety devices, warning devices, and special procedures and training (where applicable) (Raytheon 2012).

Over time, more agencies have become increasingly involved with HAZMAT regulation at higher education institutions. The NRC is the primary agency concerned with regulating radioactive materials at higher education institutions. Agencies such as the Department of Transportation (DOT) and the EPA are primarily concerned with proper waste removal for higher education institutions. A number of universities as of late have been fined heavily by the EPA for failure to remove waste properly (see Chapter 1, Figure 1.1 and Figure 1.2). Other federal agencies that have passed statutes and audited biological and chemical HAZMAT at higher education institutions include the NIH, CDC, USDA, and DEA.

There is still a bias for research universities to be focused more on reacting to HAZMAT incidents than on developing proactive measures to enhance security or safety. This is now changing to reflect the most current policies and procedures set forth by the federal government. In 2004, Oklahoma State University addressed the proper disposal and handling of HAZMAT and employee accountability for misconduct, fraud, or misuse of HAZMAT in its policy and procedures manual. However, maintaining an inventory of certain substances, protecting certain items, or contending with particular items in case of an emergency were not addressed (Oklahoma State University 1983). Currently Oklahoma State University keeps up-to-date policies and procedures on its environmental health and safety Web site that reflect the most current federal guidelines (Oklahoma State University 2012). The University of Maine addresses protection and procedural controlled usage for HAZMAT and has a provision for inventory control of controlled substances (University of Maine 2004). In 2004, however, the University of Maine did not require a centralized tracking of HAZMAT to locations and personnel. At Tulane University Health Sciences Center, the Office of Environmental Health and Safety's document titled "Hazardous Materials and Waste" in 2004 appeared more focused on proper disposal of HAZMAT than on security or inventory control (Tulane University Health Sciences Center 2004). Since 2004 Tulane has updated its policies and procedures to account for the latest federal guidelines, which is a vast improvement over the previously posted policies (Tulane University Health Sciences Center 2012). At the University of Texas at Arlington in 2004, safety was the primary focus of the policies and procedures. Inventory control was vaguely referenced "chemical inventory reporting procedures" (University of Texas at Arlington 2004). In 2012, the University of Texas at Arlington updated policies and procedures on its environmental health and safety Web site which incorporate the latest guidelines and safety standards set forth by the federal government (University of Texas at Arlington 2012). At the University of New Brunswick, policies and procedures on laboratory safety state an inventory list shall be kept on all hazardous materials. The policies also are very specific on what types of containers are used for various HAZMAT situations (University of New Brunswick 2004). Yale University has a specific section in its laboratory safety policies for lasers and centralized distribution of solvents (Yale University 2000). In 2004 Texas A&M University–Corpus Christi's policies on chemical laboratory safety focused primarily on safety issues (Texas A&M University–Corpus Christi 2004). Texas A&M University–Corpus Christi's policies in 2012 are updated to reflect the need to inventory certain items for inventory control (Texas A&M University–Corpus Christi 2012). In short, the policies in 2004 that were reviewed lack detailed information on inventory control and security procedures for use, storage, transference, or transportation of chemical elements within the university environment. Since 2012, these institutions and others have upgraded and updated their policies and procedures to incorporate the latest federal safety and security guidelines.

Many issues regarding chemical HAZMAT safety practices could be rectified by adhering to a definitive federal standard or industry guideline. However, to date no standards have been uncovered that can definitively outline how to properly secure research laboratories and HAZMAT on university property. There are several possible explanations as to why no standards can be found. One explanation is that the security measures are classified and therefore would be unavailable for public consumption. Another explanation could be that due to the great variation in research efforts being conducted, it is difficult to craft one set of criteria that can satisfy all issues that might arise in any given situation, thus forcing each organization to develop its own standards. The open access nature of college campuses would make securing most research buildings very difficult unless such buildings were located completely offsite. The age of some research facilities might prevent necessary upgrades. For example, rewiring a research laboratory that was built in the 1930s to accommodate electronic locks and surveillance cameras might be quite impossible due to the lack of available crawl space. Therefore best practices—defined as those that best fit the organizational functionality, culture, and facilities—will vary for each type of research within each type of facility. As stated by Philip P. Purpura (1989):

> Losses from crimes, fires, accidents, natural disasters and so forth, are the obvious problems. The strategies to counter these losses are numerous. A reader with even the slightest knowledge can list the more common ones: security officers, alarm systems, planning and preparation, and the like. It is of the utmost importance, however, for practitioners to look beyond these basic strategies to new methods. A never-ending search for new and better ideas is a necessity in dealing with our complex, changing world.

There is also the possibility that since securing HAZMAT is a new concern there has not yet been a centralized effort to develop operating procedures for HAZMAT issues. Therefore, it appears that the best guidelines available are the Department of Homeland Security's Chemical Facility Anti-Terrorism Standards, which not only address the need to protect HAZMAT areas and keep an accurate chemical inventory but have also been written by a federal agency with extensive influence.

In Chapter 3 we will see how the federal regulations for HAZMAT can act as an external pressure that has the ability to impact research universities on an organizational level.

3

Organizational Framework: Organizational Drift, Life Cycle, and Agency Theory

ORGANIZATIONAL THEORY WITH REGARD TO RESEARCH ACTIVITIES

In theory, research centers can be seen to rely on an "ideal type" continuum between two loci of control. At one end is the investigator-controlled research center that allows all decisions for research activities to be made by the individual researcher. At the other extreme are research activities almost totally controlled by the institution, an "organization centric" model with regard to research activity-related decisions, for example, Edgewood VX nerve gas facility in Maryland. Between these two extremes, research centers and universities' research activities are controlled, formulated, and implemented with various amounts of researcher–organizational controls. One hypothesis is that research centers and universities vary between the two points of individuality and centralized organizational control as the life cycle of the organization progresses; more specifically, as the research governance process matures, the research center or university will move to greater organizational controls given increased environmental pressures from funding agencies. With the case study university, evidence was found to support this occurrence (see Chapter 7). Other variables affecting the mix of investigator- and organizational-centered structures include the following: funding source (federal, state, or private), materials type (e.g., U-235), research design, and facility functionality for use of hazardous materials (HAZMAT).

The life cycle theory dictates that the organization will either grow or contract throughout the existence of the organization. Life cycle theorists argue that changes occur to the organizational structure over time as the organization, process, or product goes through different stages of development. The organization in its initial development has a specific set of properties. The organization in response to both external pressures and internal dynamics has its characteristics change over time—either staying in internal–external alignment, or failing to do so. Thus through time, the organization may expand, contract, or cease to exist. Bertram M. Gross (1968) describes business organizations fluctuating as the life cycle progresses by either adjusting to the economic fluctuations or by ceasing to exist.

Life cycle theorists generally argue that public organizations, not subject to market forces, are less likely to perish and more likely to undergo "renewal." A university or the research components of it, for example, would grow after initial creation and success, and contract as resources dissipate. At the time of contraction, the university will have to "renew" itself by finding more resources, new research areas and perhaps redefining its purpose. Public universities will also expand under the life cycle theory model if new directives are imposed upon the organization from an external force.

When new federal or state regulations are imposed on public institutions, universities expand organizationally, that is, devote resources, to comply with the new guidelines. Marshall W. Meyer (1979) states that environmental pressures outside public bureaus have substantial influence over expansion and contraction. In research areas, for example, a line of inquiry may fall out of disfavor with funding agencies. Internal sunk costs in laboratory equipment and personnel may not be adaptable to this environmental shift. How quickly the organization adapts to the environment is not the central issue for this research. The informing perspective is that over the course of the development of research into areas involving hazardous materials, environmental influences, especially in the form of funding agencies and their regulations, will cause aspects of the organization to change, grow, decline, and renew. Consistent funding, for example, from the National Institutes of Health (NIH) will lead the university to institutionalize NIH-specific compliance procedures beyond any single researcher.

The life cycle perspectives on organizations argue that there are predictable patterns as organizations are born, mature, and end (Daft 1995). Organizational structure in the research center or university's case will change as the life cycle progresses. These changes will dictate who the actors are and what role the actors will play in the organization. Daft (1995) has four stages outlined for progression of the organization. The entrepreneurial stage focuses on the birth of the organization where goals identified will be the focus of the actors. The collectivity stage indicates that the employees recognize the organizational goals and leadership begins to take hold of the organization. The formalization stage results in official rules, policies, and procedures being formulated and implemented by the entire organization. The final stage is the elaboration stage that results in an improvement in efficiency and effectiveness in the organization.

The life cycle perspective argues that as organizations grow, change, and contract, so do the needs and actors change and grow in power as well as the loci of control. As the organization changes, different actors will come forth to provide work on projects and provide input on how policies and procedures should be implemented or adjusted. Small research operations are less likely to evoke an administrative response than sustained large ones. Large increases in research expenditures force administrative responses to avoid organizational problems on a larger scale. If the university is increasing the amount of scientific experiments, the amount of facility space and faculty members should match the increase in research contracts and grants. As environmental factors change, the federal statutes, for instance, there should also be a need to change the organization's resources. As Herbert A. Simon (1997, 3) states:

> A great deal of behavior, and particularly the behavior of individuals within administrative organizations, is purposive—oriented towards goals or objectives. This purposiveness brings about an integration in the pattern of behavior, in the absence of which administration would be meaningless; for, if administration consists in "getting things done" by groups of people, purpose provides a principal criterion in what things are to be done.

The purpose of researchers who work with HAZMAT differs from security or emergency administrative departments that are responsible for carrying out federal, state, or local mandates. Research organizations have a vested interest in ensuring safety and security with research experiments in order to produce a result or product. The researchers in such organizations should have a particular set of skills and knowledge (in theory) to perform experimentations conducive to safety and security. The interpretation and perception of what is "safe and secure" can vary greatly between the researchers and the administration that may lack the scientific training to interpret the mandates of federal and state organizations in an applied environmental setting. Although researchers are primarily focused on how to use HAZMAT safely during experiments, they do not primarily focus on security or disposal of HAZMAT in accordance with federal, state, or local mandates. "Getting things done" for these organizations translates into being productive with research goals and objectives. Shifting focus to different objectives can cause problems for organizations.

The transition from older operational criteria to a new set of criteria can be difficult for the organization and its actors. Facets of the organization that want to maintain, for whatever reasons, the status quo will resist it. With HAZMAT issues, one problem is to implement new policies and procedures when the organization may not know or sense the urgency, saliency, or need for changes. As Kotter (1996, 2) states:

> By far the biggest mistake people make when trying to change organizations is to plunge ahead without establishing a high enough sense of urgency in fellow managers and employees. This error is fatal because transformations always fail to achieve their objectives when complacency levels are high.

From a life cycle perspective, the maturity of organizational compliance should be evidenced in the expansion of resources devoted to HAZMAT compliance in research operations to make them adhere to federal, state, or local guidelines. If the organization does not perceive HAZMAT compliance a priority, resources allocated to those efforts will decrease. Since the federal regulations have increased dramatically over the past three decades, the organization should exhibit an increase in resources allocated to HAZMAT compliance. The increase in resources for HAZMAT compliance could be stifled if the organization has other areas of growth that have been given a priority for the organization's well-being.

RESEARCHING FOR LIFE CYCLE CHARACTERISTICS

Using the life cycle perspective as an organizing framework, historical data concerning HAZMAT at the case study university, I have discovered evidence that supports the life cycle theory (see Chapter 6 and Chapter 7). By gathering the historical information on HAZMAT policies and practices, we can accomplish two objectives. The research has established how and why current policies and practices exist, and the degree to which they are in alignment with federal regulations. The research has also established recommendations to realign HAZMAT policies and practices that will improve conformity with current regulations.

Issues surrounding alignment, control, and coordination in professionally staffed, knowledge-based organizations have long been the objects of study for the social sciences (Kimberly et al. 1980). Unlike production organizations where work processes can be engineered into machines, knowledge-based organizations require a great deal of professional discretion for the work to be accomplished. Although formalized structures for performance may be in place, some social scientists argue that over time gaps between workers' performances and formal structure appear; in effect, a gap appears between the structure and worker behavior. Meyer and Rowan (1977, 356) describe this gap between informal actions and formal structure as loose coupling.

> Decoupling. Ideally, organizations built around efficiency attempt to maintain close alignments between structures and activities. Conformity is enforced through inspection, output quality is continually monitored, the efficiency of various units is evaluated, and the various goals are unified and coordinated. But a policy of close alignment in institutionalized organizations merely makes public a record of inefficiency and inconsistency.

Decoupling as a concept has a close theoretical relationship to the concept of organizational drift. Both concepts point to a problem for the organization. The organizations' actors are performing a set of actions that are, over time, less and less related to what is needed. Perrow (1999) states loosely coupled systems have the flexibility to incorporate change (e.g., failure, shocks, etc.) without destabilization. However, loosely coupled systems also do not react quickly to required change, leaving the organization potentially vulnerable. Although tightly coupled systems will react quickly to change, the organization could also overreact, with dire consequences (Perrow 1999). Universities vary on how tight coupling occurs in the organization in relation to regulatory agencies changing guidelines. Some universities will be more responsive to regulatory changes than other universities due to the composition of the faculty and staff at the institution.

Drift can occur for a variety of reasons, but relevant to this research it can occur because of an "indifferent regulatory environment." That is, regulatory agencies assume that researchers and their larger university environment are conforming to their regulations. Meyer and Rowan (1977) describe this assumption as the "logic of confidence," a kind of good faith in organizations and their professionals. As stated by Meyer and Rowan:

> Despite the lack of coordination and control, decoupled organizations are not anarchies. Day-to-day activities proceed in an orderly fashion. What legitimates institutionalized organizations, enabling them to appear useful in spite of the lack of technical validation, is the confidence and good faith of their internal participants and their external constituents. (p. 357)

If a funding agency provides research funds to a researcher at a university, it is often assumed that the university has in place and implemented a set of controls for risk and security, and that the researcher is complying with these controls. Thus, there is no reason for the agency to look closely at what might be occurring. Likewise, the administrative apparatus of the university can assume the researcher is complying with policy. Hence, rippling through the funding system can be a "logic of confidence" that each element is in fact performing as it is assumed to be performing. Since in theory no one is looking, there can be "drift" at multiple levels, from when the researcher establishes procedures and policies to oversight and audit schedules by the regulatory agency. Purpura (1989) illustrates how internal and external factors can affect organizational structure and protection programs that will lead to drift.

> Management theorists often study factors that influence organizational structure. These factors provide a foundation for understanding why enterprises are designed in certain ways. Both internal and external factors affect structure and have a bearing on protection programs. The astute security and loss prevention executive perceives these factors and responds to the needs of the entity. (p. 72)

Whatever the security aspects are for an institution, internal existing organizational structure will need to be considered when formulating proper procedures or policy to secure assets such as HAZMAT. Friction points can exist between the upper administration and the administration of each individual department. Miller (1998) states that bureaucratic dysfunction arises when various subunits pursue their own interests within their sphere of operation.

At a university, research faculty consists of highly educated individuals who work with a wide variety of assets to accomplish research goals. Purpura (1989) argues that differences in educational backgrounds lead to differences in the efficiency of the organizational structure, with more educated workers preferring a participatory style and less educated workers feeling more comfortable with the bureaucratic style. If Purpura's assertions are true, researchers should expect (if not demand) a more participatory style of management. As stated by Kammen and Hassenzahl (1999, 11):

> Proponents of a participatory philosophy argue that risk analysis remains too subjective, and its implications too dependent on social context, to permit its removal from the public arena. Since decisions about values and preferences are made not just at the final decision stages but throughout the process, risk analysis necessarily combines both technical expertise and value choices. The implications of this interplay range from the inadvertent, as analysts make choices they believe are best without input from interested parties, to the antidemocratic, when the value decisions as well as the number crunching are intentionally restricted to a select group with a particular agenda.

When devising a security methodology, one should take into account the possibility of research faculty resisting new bureaucratic rules that could be perceived as hindering productivity. With regard to HAZMAT, various federal agency guidelines mandate controls be in place in the organization so that productivity will not be degraded through loss of materials, destruction, injury, or death. From "loose coupling" perspectives of organizations, then, the degree to which policies and practices are tightly followed in practice bears empirical examination. Perrow (1999) states the potential for disaster of DNA research upon society if tight procedures and policies are not in use due to interactions between systems that were not linked (or not foreseen to be linked) can lead to epidemic proportions. One possible mechanism for decoupling is the pursuit of individuals' interests over the interest of the organization as a whole. This mechanism has been recast as asking "On whose behalf or as whose agent is the actor acting?" Researchers and theorists interested in this question have captured their arguments within the rubric of "agency theory," which often focuses on the behavior of different types of actors within public bureaus. According to Downs (1967), the beginning of bureau life cycles having three similar traits of rapid growth, domination by zealots or advocates and the seeking of resources for survival.

An "agency theorist" viewing researchers might argue along the following lines. All research organizations have a research mission to create new and substantial knowledge about our world. The pressures to accomplish this mission fall squarely on the researchers. Researchers are dual actors. They act both as agents of the organizations and as their own agents in the larger arena of their scientific endeavors. Researchers control a bundle of resources: laboratory space, specialized contracts and grants services, research assistants, special access to new equipment funds, and so on. The research team or its agents, in effect, become advocates for a particular kind of organization of units that intersect with them. From a strict "agency theory" point of view, these agents would strive to have arrangements that would maximize the creation of new resources for them and minimize the transaction costs to doing so. This may be especially so since success—both within the research part of the organization and in the scientific arena—is often measured by extramural funding that the faculty attracts. As Cyert (1978) notes, this revenue stream provides higher education faculty in particular with a great deal of independence: "The faculty member not only has independence in the classroom but also gains additional independence because he is a source of revenue to the organization through his research efforts (p. 344)." This independence, while unique among organizations, is generic in research universities. The independence and greater workplace freedoms that faculty are allotted is a factor in how universities are organized. It also deeply influences how rules concerning research are made and enforced. Researchers, in effect, can establish and maintain policies and procedures for research that became practices for the organization. There may or may not be administrative review or oversight. Under what conditions, then, is there a change in the loci of control? Cyert argues:

> When the organization is functioning effectively, as measured along educational, research, and fiscal dimensions, the structure may present little difficulty because only limited central action is necessary. The situation is quite different when a contraction in size is forced on the organization, and it becomes clear that the central administration must take action. The contracting process brings with it conflicts of all kinds in contrast with an expanding organization, p. 344.

Cyert (1978) notes one (of many) conditions that cause central administrations to take a more aggressive role in governance. A change in the external regulatory environment surrounding universities (or research centers for that matter) in so far as regulatory agencies become more intrusive and demand greater accountability from the university can alter the governance of research activities. When granting agencies—federal or state—create new regulations, they assume research centers and researchers will make the necessary alterations in policy and behavior to be in compliance. Thus, over time, policy and behavioral effects should be in evidence. Research activity in a faculty lab, once the purview of the individual faculty and operated solely under traditional scientific mores, may become subject to externally imposed regulatory conditions and compliance checks. Funding and regulatory agencies may desire greater accountability in all domains of the research process: lab space utilized, faculty times, materials consumption, safety procedures, and hazardous and regulated materials safety and use. As more compliance becomes necessary, the central administration is likely to take steps to ensure compliance by individual faculty for the sake of the university as a whole.* This tension between centralization of compliance and individual researcher autonomy is endemic in modern research organizations. For example, Perrow (1999) states that researchers are against formulating stricter guidelines on DNA research due to fear of being sanctioned by institutions for raising differing points of view.

From the agency point of view, researchers make calculated decisions that will benefit their position for increasing productivity in research. The policies and procedures faculty formulate and implement may or may not be in compliance with the new regulatory environment. In the "loose coupling" scenario, it is presumed they know what they are supposed to be doing and are doing it; on the other hand, as noted in a later section, evidence suggests that in some instances this is not the case. Monroe (1991) states that individuals make decisions based on certain expected outcomes or goals that advance the agent's agenda in a rational manner.† Agency theorists might argue that if regulatory compliance is seen as

* Life cycle theorists also argue the amount of centralized control varies by the stage of the "life cycle" the organization finds itself in. During the entrepreneurial stage of the organization, for instance, the amount of central control will be minimal and the researchers will be the prime agents in the organization. However, as the organization moves to the formalization stage, central control will increase. See the example in Daft (1995).

† As stated by David Johnston in *Human Agency and Rational Action*, p. 94.

slowing scientific creation or interfering with established procedures, the agent with large autonomy may be reluctant to implement change. This might seem to be especially the case when the central administration is confident that the researchers are in compliance.

Universities, as organizations, encourage research faculty to be innovative and productive in obtaining grants and research activities. Research universities benefit from faculty success by gaining prominence in a field of study and by increasing the potential of appropriate funding from donations, contracts, and grants. The more productive a research faculty member becomes, the more revenue is generated for the university. Theoretically, based on the "Los Alamos model" of imposing too many restrictive social controls ("policing your employees by every means of procedures you can devise") will interfere with the research process and can degrade this productivity. At some point, increased procedures will deter creativity within the organization (Morgan 1989). At a university, faculty members who are not creative are not productive. Thus, introducing social controls to police every detail can be counterproductive to producing research at the organizational level. As stated by Herbert A. Simon (1997, 35–36):

> In a large organization with interrelations between members, a restricted span of control inevitably produces excessive red tape, for each contact between organization members must be carried upward until a common superior is found. If the organization is at all large, this will involve carrying all such matters upward through several levels of officials for decision, and then downward again in the form of orders and instructions—a cumbersome and time-consuming process.

A collaborated effort between research faculty and administration to design social controls is preferable for enhancing security and productivity. Increasingly, work is completed by a horizontal process, else the work runs the risk of remaining incomplete since a wide array of entities does not have a stake in the process (Morgan 1989).

If the researchers do not have input or vested interest in following formulated social controls, then the social controls risk being ignored and not used. Hall (1972) states that the greater dependence an organization has on research will cause the organization to have lower controls and higher professional incentives. As agency theory indicates, each entity will act according to its individual agenda as opposed to the organizational agenda. Without accountability in an organization, agents are allowed to drift to their own agendas, and in theory allow policies and procedures to degrade over time to expose the organization to risk and hazards.

The theoretical perspectives briefly described inform the research. Each applies to a different unit of analysis in organizational structure and process. Agency theory focuses on the motivations of the actors in the organization from a rational model. Life cycle theory sensitizes the researcher to attend to the organization's structure and mission over the life course of the organization with special reference to environmental change and pressures. Coupling and drift perspectives alert us to assumptions made by actors and to the empirical analysis of the conditions under which policies and procedure may or may not be translated into organizational actions. This leads to the discussion in Chapter 4 on how data is collected to assess the organizational theories in relation to the HAZMAT situation at the University.

4

Threat Matrix to Public Research Organizations

SAFETY AND SECURITY ISSUES FOR RESEARCH ORGANIZATIONS

Providing security within an environment that fosters the free exchange of research information among internal and external professionals among a population of transient students can be daunting. Such an environment can make relatively simple tasks like the identification of faculty and staff and the tracking of assets used for various research purposes among multiple departments more complex. Maintaining adequate security at a research university can be difficult due to the physical size of the facility and the vast number of people it serves compared to the size of a typical university police department. Students are frequently found in facilities after hours in restricted areas. Law enforcement personnel often cannot determine which individuals are students with legitimate access and which are trespassers. "Assets such as intellectual property can be more valuable than physical assets" ("With Some Strings Attached" 2003). Thus, it is important to monitor who has access to what areas of the institution. A committee at the Massachusetts Institute of Technology (MIT) addressed this issue by recommending that laboratories engaged in classified research be relocated off-campus to ensure tight security over those endeavors while maintaining open access on the main campus (Campbell 2002).

Universities that have active research programs often store hazardous materials (HAZMAT) in their facilities. This necessitates strong security measures that will guard research projects against theft and the improper use of chemicals, biological specimens, and radiological materials. As stated by Nolan (1996):

> Human factors and ergonomics play a key role in the prevention of accidents. Some theories attribute up to 90% of all accidents are caused by human factor features. It is therefore imperative that an examination of human factors and ergonomics be undertaken to prevent fire and explosions at petroleum facilities since historical experience have also show it is a major contributor either as a primary or underlying cause, p.240.

Law enforcement officials who enter research laboratories during an emergency situation risk exposure to dangerous HAZMAT materials that could cause a fatality, especially if the officers are unaware of the presence of such materials. Since many campus security forces work with limited resources and labor shortages exacerbated by budget constraints, it begs the question how an institution maintains the safety of its facilities and personnel when the policies, procedures, and tools that can enhance law enforcement's existing capabilities do not exist. This brings up three issues for law enforcement: identifying the location of HAZMAT, monitoring HAZMAT, and establishing protocols in the event of the theft of or an emergency situation with HAZMAT. In reference to safety and security of biospecimens Hawley and Eitzen state:

> Safety and security can be achieved by controlling access of personnel to the laboratory area(s). Access through a locked door can be controlled with a key-operated lock or with an electronically controlled card that is keyed to the identity of the individual and, if applicable, to their safety training and immunization status. ... Safety and security could also be enhanced by requiring approval of the institute director, or his or her designee, before an investigator may use and study an etiologic agent. (As quoted in Fleming and Hunt 2000, 572)

The federal government is aware of the importance of higher education institutions controlling HAZMAT situations. As stated by the U.S. Environmental Protection Agency (2004a):

> The EMS Work Group (Emergency Medical Services) is currently working to develop tools and resources necessary to promote EMSs on campuses. The Group is focusing on strategies to address colleges and universities: (A) with no previous consideration of implementing an EMS; (B) those that have made EMS considerations, but require assistance; and (C) those that have already begun the EMS development and implementation process but could benefit from specific and more quantitative EMS strategies.

Institutions have a duty to ensure that hazardous materials do not fall into the possession of terrorists or criminals. The United States Department of Health and Human Services recently cited eleven unnamed research university laboratories for having inadequate physical security measures for laboratory pathogens or deadly toxins (Field 2004). According to Field (2004):

> Universities have begun abiding by a new set of safety regulations issued by the Agriculture and of Health and Human Services. Those rules require labs to register with the federal government, disclose their inventories of "select agents," and establish security plans. Select agents are identified in federal regulations as micro-organisms or infectious substances capable of posing severe threat to public health and safety.

No matter how thorough the security or protective measures are, the effectiveness of these measures relies completely upon the willingness of the personnel to implement them. As stated by the National Research Council (1991):

> A chain is only as strong as its weakest link. If one wishes to carry a heavy burden, the strength of all links in the chain must be at least adequate for the load, and there is little value in having a few stronger links if the weak ones break. And so it is with the safety of buildings.

Faculty can affect the implementation of new security measures. As described by Jeff Schiller, network manager at MIT, during an interview with *Syllabus* ("A Balancing Act?" 2004) on network security:

> Speaking as a network manager at an institution with Nobel laureates, it's harder for me to set policy and make it stick. The more famous your faculty, the more they're in charge. And the more the faculty can do whatever they want, the more chaotic your network's going to be.

If personnel do not accept enhanced measures and choose not to use new policies, procedures, or methodologies, then the risks of crisis will be similar to not having any protection at all.

Tightened budgets and the increased cost of operations have produced a scarcity of resources in today's public higher education institutions (Schwartz 1987). Consequently, there is an increased need to optimize resources as a way of significantly reducing deficits, waste, and institutional vulnerability. The primary fixed assets of public higher education institutions are their facilities and the land on which they reside. Institutions, therefore, need to ensure that effective policies and tools are in place to safeguard existing infrastructure. As stated by William P. Hoye (2004): "In the final analysis, the best institutional hedge against increased legal liability is improved risk identification, risk assessment and risk management with respect to the university activities and programs."

Awareness of the various threat categories can enhance security efforts at research universities without requiring large increases in the security budget.

HIGHER EDUCATION THREAT CATEGORIES

There are six primary categories that relate to campus security and allow for proactive measures. Category I is a high threat as it pertains to organized criminal activity. Category II is a high threat as it pertains to individual felony activity. Category III is a high threat due to natural disaster. Category IV is a high threat of an accidental nature. Category V is a medium threat. Finally, category VI is a low threat due to misdemeanor activity (see Figure 4.1).

To conduct this research, it is necessary to answer several basic questions. Which of these categories cause the most damage? What incidents are preventable through effective policies, procedures, or tools? What can

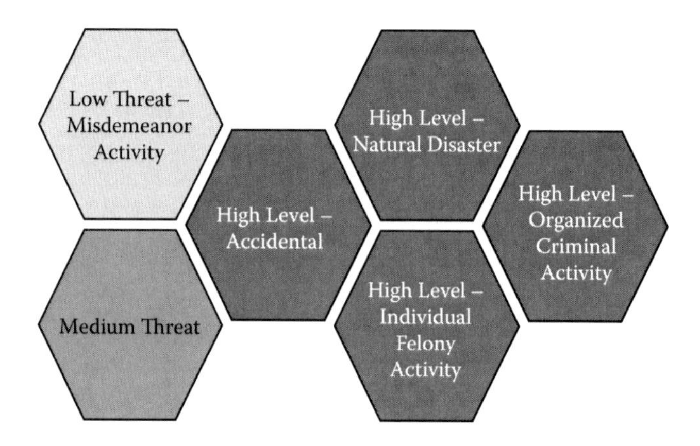

FIGURE 4.1
Honeycomb chart for higher education threat categories.

institutions do to increase their security level? What upgrades are necessary to respond to the threats outlined in each of the categories?

Categories I and II

Since 9/11, terrorist actions have been the focus of law enforcement's preventative and proactive efforts. High-threat categories I and II deal with terrorism as well as traditional organized criminal activity. In the past, some institutions have dealt with high-threat categories that include damage from bombings, mass murder, terrorist activity, or similarly severe actions.

The bombing of university facilities has become a major issue over the last forty years. In the 1960s, a terrorist group named the Weatherman planted a bomb in the Harvard Center for International Affairs (Lundegaard 2003). The Unabomber is another example of terrorist actions that resulted in damage to several university buildings and serious injury to staff and faculty. The Unabomber targeted the University of Chicago (1978), Rensselaer Polytechnic Institute (1997), Northwestern University (1997), University of Utah (1981, device neutralized), Vanderbilt University (1982), University of California at Berkeley (1982 and 1985), University of Michigan (1985), University of California at San Francisco (1993), and Yale University (1993) (Ottley 2004). Ten universities had highly explosive bombs placed in their facilities. Most of the devices looked like legitimate mail parcels. Scanning and opening every piece of mail would prevent bombings at university facilities. However, this does not mean that preventing a bombing is practical or feasible for the average security force at a higher education institution. Protecting the mail delivery system was required in order to defeat some of the Unabomber attacks.

Murder, terrorism, and hostage-taking are unfortunately genuine concerns on college campuses given the large numbers of people who walk throughout the facilities at all times of the day and night and, in many cases, unmindful of their surroundings. It can be a simple matter for a mass murderer or a terrorist organization to inflict harm on large groups of people with great speed from hidden vantage points. As in the case of Charles Whitman at the University of Texas at Austin, the clock tower provided Whitman with a wide field of fire and ample protection from law enforcement (MacLeod 2004). Although the death count could have been much higher if Whitman had begun his rampage during a time when students were changing classes, this incident could have been prevented if access to the tower had been restricted. Another example is Ted Bundy's attack on the sorority house at Florida State University (Michaud and Aynesworth 2004). Due to a lack of surveillance equipment and an inadequate security force, Bundy was able to sneak into the sorority house and commit atrocities on young women. At Case Western Reserve University, a former employee killed one student and injured another student and a professor as part of a seven-hour attack (Hakim 2003). The police were eventually able to subdue the man by shooting him in the shoulder and abdomen (Hakim 2003). In 2010 six faculty members were shot of which three of them died at the University of Alabama–Huntsville during a faculty meeting allegedly by a faculty member who had been denied tenure (Bartlett and Wilson 2010). These incidents highlight the need to protect campus residents not only from outside threats but also from threats within the university community.

Kidnappings and assassinations of very important persons (VIPs) can occur from either organized or nonorganized criminal activity and thus can be listed under either threat category I or II. For example, the Unabomber targeted specific well-known research scientists in fields such as genetic research, computer science, and engineering. Hosting VIPs requires an increase in security and cooperation with a wide array of law enforcement agencies. The case study university has frequently hosted VIPs for press conferences on a variety of topics. Past visitors to the case study university have included United States presidential nominee (later president) George W. Bush and the president elect of Mexico (later president) Vicente Fox. To illustrate the point, during President Obama's inauguration, several facilities were impacted by the Secret Service as they prepared for the VIP. As stated by Purtell (2009, 24):

> Imagine this scenario: anywhere from one to four million people are predicted to pour into your medium-sized city for an historic presidential inauguration, the Secret Service has the city in lockdown mode, and your building happens to sit directly in the parade route. Now add another minor wrinkle to this scenario: people in your facility, several of them in fact, want access to the building to throw inauguration bashes. Securing your building and satisfying these people's needs may seem like mission impossible in such a situation, but fms [facilities managers] in DC had to do exactly that.

The theft or detonation of nuclear, biological, or chemical (NBC) specimens is a relatively new threat that can cause widespread damage to persons and facilities. Most higher education institutions own NBC materials for research purposes or as cleaning compounds. Illegal entities, such as terrorist groups, seek NBC materials to make weapons or make a profit on illegal arms sales. The Aum cult is a prime example of an entity obtaining a biological compound for their NBC development program (Bellamy 2004). The Aum obtained one sample of anthrax through one of their members who was a graduate student with a medical contact in the bioweapon development field (Bellamy 2004). The Aum also conducted a sarin gas attack in the Tokyo subway system, proving that chemical attacks can be easily launched against large masses of people with deadly consequences (Bellamy 2004). Another example of a successful biological attack was the anthrax attack in 2001 via the United States Postal Service. The attack revealed how vulnerable the U.S. mail system was to terrorist attacks. Preventing the illegal use of an institution's NBC materials is achieved by ensuring that areas that store NBC materials are tightly controlled and patrolled. Preventing external attacks requires a larger labor force and more tools. However, even these measures may not be enough to stop a determined group that is well organized and ruthlessly intent on attacking a facility or a public area.

It is very difficult to prevent a determined attacker from igniting a car bomb or engaging in a suicide bombing on institutional grounds. Unless security forces are coincidentally in the area or have received specific warnings of an imminent attack, these types of attacks can occur swiftly and can inflict severe damage. In the case of car bombings, physical barriers can prevent a car from getting too close to a facility, but almost every building has some type of loading zone where a vehicle can get close to the entrances for service delivery. The Oklahoma City Federal Building (1995) and the Beirut United States Marine Corps barracks (1983) bombings illustrate how much damage a car bomb can cause (Slavin 2004). In the case of suicide bombers, no security system can stop this type of activity short of frisking every person on campus. This is not a feasible solution for higher education institutions.

The theft of intellectual property and other assets by researchers, graduate students, or staff members who have obtained security clearance, although not preventable, can be made more difficult with security measures. Through constant vigilance and security upgrades, information technology departments can prevent hackers and viruses from attacking university infrastructures. However, unless security personnel follow all employees and students, tap their phones, and regularly check their hard drives, it is impractical to prevent the theft of intellectual property. Physical forces can prevent more blatant efforts at intellectual or asset theft. As demonstrated by a recent occurrence at the Los Alamos National Laboratory, physical forces stopped two Chinese diplomats who ran past a checkpoint on a secure road near a plutonium research facility (Rankin 2004). The diplomats were eventually escorted off the property. Since institutions have limited security forces and modest resources, setting security priorities is necessary to ensure that certain items that are critical for institutional operations are secure. An officer's handgun is required for security operations and can be used by assailants to compromise campus security. However, classroom supplies, such as a stapler, are far less critical. Precautions must exist to protect the officer's sidearm in the arsenal even if it means compromising the security of less critical assets, such as classroom supplies.

Embezzlement is almost always an internal threat to a university. Security forces can only react to reported embezzlement. The institution itself is responsible for setting policies and procedures that can lessen the probability of embezzlement or lead to the swift discovery of embezzlement activity, such as the use of background screenings of new employees. However, even with these measures, embezzlement can occur even in places where security measures in other areas are quite sound. In 2004, two former workers for Los Alamos National Laboratory were indicted on embezzlement, fraud, and conspiracy charges (Benke 2004). A former university employee of the University of Texas at Brownville and Texas Southmost College and another suspect were sought by authorities in a series of incidents that occurred at three different universities and colleges where computers and electronics were stolen. One suspect was apparently caught on surveillance videotape entering a room without a computer and later leaving the room with a computer in a backpack (Associated Press 2004b). Former employees are difficult to apprehend because they have knowledge of the environment, the security infrastructure, and where valuables are typically stored.

Civil rights concerns coupled with heavy labor demands make the prevention of drug crimes, assaults, armed robberies, arsons, rapes, and other such crimes exceedingly difficult. These are crimes of opportunity and are prevented only when security forces notice suspicious activity while on patrol. Therefore, removing opportunities for criminals to act can lower the chances of such crimes occurring. Surveillance equipment is a useful tool that can assist officers in identifying suspicious behavior before a criminal act occurs. Proper lighting and visible security measures in the form of cameras or officers on foot patrol or in squad cars can thwart armed robbers and rapists who require isolated, dimly lit areas to commit their crimes. Simply ensuring that visitors, students, staff, and faculty are not in the facilities after normal business hours unless they have special clearance can prevent many crimes.

With some crimes, such as illegal drug activity, threats, or trespassing, vigilance and a willingness to take all potential security risks seriously is all that enables campus police to prevent more serious occurrences from happening. Illicit drug activity can occur for weeks without security personnel being aware of it. Typically, violent crime associated with illegal drug trafficking, a witness's information, or a drug overdose is often the first sign that drug activity is occurring. Recently at New York University (NYU), a student managed to live in the university library for over eight months despite being seen on occasion by security personnel (Matthews 2004). Although this student was benign, this incident highlighted the fact that students could have engaged in criminal activity for months because security personnel did not consider them a credible threat to campus security nor considered the library a high-risk area. Conversely in 2004, police took a professor at the University of Louisiana at Lafayette campus into custody after he allegedly threatened to kill his students (Fogg 2004). Although this instructor never committed an assault, security took his verbal threats seriously and quite possible prevented injury or death through proactive engagement of the potential threat.

With continued budget reductions, campus security forces need to become more flexible in their response to problems that can spontaneously erupt in a variety of areas. Many of these incidents are preventable by being able to identify potential assailants and by securing potential human targets.

Category III

Natural disasters are not preventable. A hurricane will either strike the coast where a university resides or the hurricane will change its path at the last minute. However, universities can mitigate damage from natural disasters through engineering, sound policies, or the use of strong containers for HAZMAT. The University of Texas Health Science Center in

Houston had several laboratory animals and research projects destroyed when the laboratories were flooded (Bret Wandel, asset coordinator at UT-Health Science Center Houston, personal communication 2004). These facilities were located in the basement. If policies had existed that required such laboratories be located above ground, the flood damage might have been avoided. However, in other instances, the location of the campus itself is a liability and one that policies cannot mitigate. In 1993, the University of Iowa campus was flooded when rainfall exceeded the nearby reservoir's capacity (Moninger 2003). East Carolina University also experienced flooding and other damage when Hurricane Floyd struck the North Carolina coast in 1999. One student drowned and the campus at Greenville had damage that exceeded $7,000,000 (Corbett 1999). The storm also hurt the surrounding community by damaging or destroying many homes. Hurricane Wilma in 2005 caused damaged to Lynn University with 120 mile per hour winds and flooding (Villano 2009). Natural disasters can cause a significant amount of death, damage, or injury to the campus population, to the institution's assets, and to the surrounding community. During such an event, sometimes the only thing an institution can do is to prepare emergency medical services (EMS) and security teams to respond quickly and effectively when the event occurs.

Category IV

Occasionally, accidents do occur at university facilities. More often than not, accidents happen when either policies or procedures do not exist or when people choose not to follow them. In 2003, Texas Tech University notified the Federal Bureau of Investigation that several vials of *Y. pestis* cultures, also known as bubonic plague, appeared to be missing. What initially appeared to be a theft was later revealed to be an accident due to the fact that the scientist who was responsible for the cultures did not maintain a careful inventory (Eiserink and Malakoff 2003). The university did not provide adequate oversight of his activities and so this accident was not detected until the cultures were missing. Although absolved of some of the federal charges, the scientist was convicted of many of the fraud charges and of charges of illegally exporting materials (Eiserink and Malakoff 2003). To this day, thirty vials of *Y. pestis* are still missing. The Texas Tech case illustrates how one issue can cut across multiple threat categories. The lack of oversight and human error could have been factors in the Texas Tech situation.

Higher education institutions are not the only ones to experience mishaps with biological specimens. In 2001 and 2002, the U.S. Army Medical Research Institute for Infectious Diseases (USAMRIID) laboratory had a series of anthrax contaminations from samples that had escaped the laboratory. Some of these samples were supposed to be secured in a laboratory designated as a biosafety level 3 (Vergano and Sternberg 2004). It was determined that the contamination occurred after the number of laboratory personnel was increased from six to eighty-five members, some of whom had to learn on the job and had no previous experience dealing with bacteria. Their lack of experience and training led to the accidental contamination of various areas.

Seton Hall is an example wherein both criminal acts and negligence played a major role in causing death and injury to students and damage to a building on January 19, 2000 (Cohen 2003). As stated by Cohen "About 1,600 fires break out in dormitories and fraternity and sorority houses each year, with 65 percent of those taking places in facilities without sprinkler systems, according to the National Fire Prevention Association" (Cohen 2003). Arson was determined to be the cause of the fire, and the authorities arrested two students for setting the blaze. However, Seton Hall had no sprinkler system installed in Boland Hall. Three students died and 58 students were injured. Seton Hall is not alone in contending with fires in dormitory facilities. In recent years fires have claimed lives at Bloomsburg University in Pennsylvania, the University of Pittsburgh in Pennsylvania, the University of California at Berkeley, the State University of New York at Binghamton, New York University, Millikin University in Illinois, the University of Dayton in Ohio, the Massachusetts Institute of Technology, and Lee College in Tennessee (Cohen 2003). Although arson falls into a more serious threat category, fire accidents can be avoided with proactive measures such as sprinkler systems, personnel training, and so forth. Arson was also responsible for a June 3, 2008, incident when the Animal Liberation Front burned an empty van belonging to the University of California at Los Angeles that was used to transport faculty members (Vinas 2008).

Universities are contending with a new form of "accident." Vulnerable databases, networks, and other information resources are now the targets of hackers and high-tech thieves. In 2003, hackers obtained the names and social security numbers of 59,000 current and former students, faculty, and staff from the computers at the University of Texas at Austin

(Haurwitz 2003). Obviously, the university's information technology department did not provide adequate security for their server.

> "We flat out messed up on this one," said Dan Updegrove, the university's vice president for information technology. "Shame on us for leaving the door open, and shame on them for exploiting it. Our number one goal is to get those data back before they get misused." (Haurwitz 2003)

With the increased use of computers for administration and research, institutions need to increase their vigilance in the information resource area.

When personnel choose not to follow procedures, the consequences can be dire. In August 2003, two workers at Los Alamos were accidentally exposed to radiation that leaked from corroded containers (Wright 2004). This accident was the result of negligence by personnel who did not keep radioactive material in proper storage containers. The National Nuclear Security Administration (NNSA) found that Los Alamos had violated seven safety regulations (Wright 2004). In 2003, the northeastern portion of the United States witnessed a widespread power outage that lasted for days. Both United States and Canadian officials reviewed the incident and discovered that the event was preventable. First Energy Corporation in Ohio did not follow industry policies and procedures. This failure allowed the blackout to occur (Perez-Pena 2004).

Category V

Higher education institutions have witnessed several riots and protests over the last forty years. Kent State University (KSU) is a prime example of a campus where riots and protests erupted during an anti-Vietnam War rally. On May 2, 1970, protesters torched KSU's ROTC building. According to the Federal Bureau of Investigation (FBI) report, one of the leaders of the protest was a KSU professor who threatened students to engage in the activity or suffer retribution by receiving lower grades. As the firefighters attempted to extinguish the fire, protesters damaged the firefighting equipment. The Ohio National Guard was brought in to control the riot. In the ensuing struggle, the Ohio National Guard injured one student with a bayonet. On May 4, National Guard troops fired into groups of students who were protesting. The shooting resulted in the death of four students and the injury of nine additional students (Federal Bureau of Investigation 1970).

During this riot, incidents from more than one threat category took place. The students committed arson and the troops committed homicide. Policies and procedures to effectively deal with protesters were apparently lacking. Campus security forces could have protected campus grounds with local law enforcement instead of utilizing Ohio National Guard troops. The fact that KSU continued classes during the riots illustrates a lack of foresight in assessing the possible injures that could occur if the situation became uncontrollable. Since there were more people in the center of campus going to and from class, this presented more opportunities for death or injury from National Guard troops.

Other crimes, such as drug use, vehicle theft, and property theft, can be deterred with increased patrols and surveillance equipment. Recovery of stolen vehicles or property may be possible if surveillance equipment has successfully recorded the suspect's image for identification.

Category VI

Although misdemeanor activities are more of a nuisance than a threat, these offenses can lead to the discovery of more serious offenses. For example, if a police officer pulls over a suspect for speeding on the institution's road, the officer may find explosives in the vehicle intended for destruction of a facility. Category VI offenses require security forces to address these concerns in order for an institution to maintain authority over its property. For example, if a vandal disfigures a building with spray paint and security does not address the issue, little would deter that vandal from defacing another building or attempting a more serious crime such as theft or arson. Some offenses cut liability and reduce the risk of injury for students, for example, enforcement of traffic laws. Category VI threats are reactionary responses rather than proactive measures; however, security forces still require resources to address these types of infractions.

ON WHICH CATEGORIES TO FOCUS AND WHY

This research focuses on the ability of physical forces to act proactively to prevent certain dangers or mitigate damage, injury, or death, and will explore new measures that will assist in the apprehension and prosecution of criminals and the recovery of stolen items (Figure 4.2).

FIGURE 4.2
Areas of focus for physical security forces.

The use of surveillance cameras can deter crime as well as assist security forces in the apprehension and conviction of criminals and the recovery of stolen items. In some instances, cameras may only be able to assist security forces with the aftermath of a crime. Feasibility and practicality are required for security forces to use new policies, procedures, and tools, and for those tools to be effective.

For example, it is not feasible to inspect every piece of mail that enters the university mail system for chemicals, biotoxins, or explosives. However, it may be feasible to have a policy to inspect every package that is returned with an incorrect mailing address. By creating a quad threat chart (Figure 4.3), physical security research and methodologies will focus only on events that physical security can affect.

The research will recommend policies and procedures for areas of concern that physical security forces do not influence, for example, criminal background checks as part of hiring practices. For example, in the "Prevention" area of the quad chart, physical security action could have prevented the loss of plague samples by inventorying the contents of the professor's lab. In the "Mitigation" area of the quad chart, a checkpoint at the base of the University of Texas at Austin's tower might have prevented Charles Whitman from using the tallest point on the university as a sniper's nest and, thus, would have lessened the fatalities committed.

In the "Apprehension" area of the quad chart, proper surveillance equipment might have apprehended the Unabomber sooner if cameras recorded his movements at universities where he hand-delivered his bombs. However, as pointed out in the quad chart category "Not Impacted by Physical Security Measures," physical security forces will not affect a hurricane or espionage. Some of the given examples are appropriate for one or

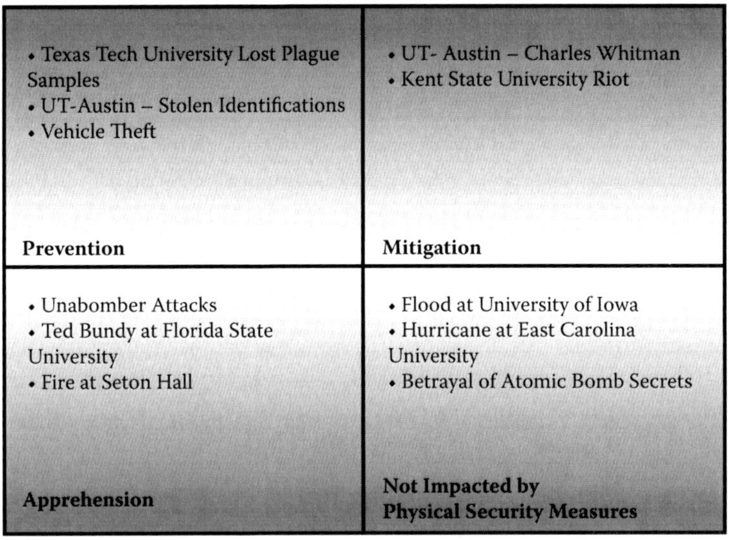

• Texas Tech University Lost Plague Samples • UT-Austin – Stolen Identifications • Vehicle Theft **Prevention**	• UT- Austin – Charles Whitman • Kent State University Riot **Mitigation**
• Unabomber Attacks • Ted Bundy at Florida State University • Fire at Seton Hall **Apprehension**	• Flood at University of Iowa • Hurricane at East Carolina University • Betrayal of Atomic Bomb Secrets **Not Impacted by Physical Security Measures**

FIGURE 4.3
Quad chart of physical security impact zones.

more areas of the quad chart. For example, a proper fire sprinkler system could have prevented the Seton Hall fire. Surveillance equipment in dorm hallways would also assist security forces in the quick apprehension of arson suspects.

5

Methodology

INTRODUCTION TO RESEARCH DESIGN

The focus of the research is to gather field data on organizational actions at the University that relate to the safety and security of hazardous materials (HAZMAT), and to gather policies and procedures illustrating practices from a selection of other universities. Since locating archival records and other documentation was difficult, multiple methods were used in data collection. At the University, much of the institutional history and current practices for HAZMAT are not documented. Instead, this information is contained within the collective memories of personnel, both retired and active, who remember certain practices, the evolutionary history of the institution, and the status of any documentation related to the topic. Sets of observations on HAZMAT were gathered that enabled the use of grounded theory with the research. During the data collection phase of the research, ethical dilemmas often presented themselves throughout the process, aspects of which will be discussed later in the chapter.

Since the data were located in various places, multiple techniques were used to extract the data for use in the research. Each technique used for the collection of information presented significant challenges. Face-to-face interviews were helpful for collecting information not found in other sources of data but had the disadvantage of forcing the researcher to rely on the respondent's memory. Archival documentation was useful for establishing a timeline of events and for uncovering data on organizational operations in relation to HAZMAT. The disadvantages of using archival information is that the context of the documents themselves may be interpreted differently, documentation may not exist for certain practices or procedures, or the documentation may exist but is not utilized by the organization. Unobtrusive observation is a useful method for collecting data on ongoing organizational activities. However, this method may provide incomplete or distorted information because the moment in time in which the data was collected may contain conditions that do not replicate themselves at another point in time. The researcher as a participant observer has the advantage of gathering data that by normal methods would not be collected, but there is a danger that the researcher can become emotionally affected and thus be biased in his or her conclusions. By using several different methodologies, a researcher can triangulate data from many different sources to attain a relatively reliable source of data for analysis, recommendations, and conclusions.

DATA COLLECTION

The case study university will be referred to as "the University"; moreover, the forerunner of the University will be referred to as the "Research Think Tank" throughout the research. The names of interviewed respondents have also been withheld for confidentiality reasons. The research methods used to gather information on the following topics will be discussed in separate sections.

- The University's history
- The University's current practices on contending with HAZMAT
- Current University HAZMAT safety and security polices and procedures
- Federal research statutes and guidelines with regard to universities and HAZMAT
- Best practices and institutional organizational structure for universities that contend with HAZMAT issues
- Federal research investigations for universities and HAZMAT regulations regarding research activities
- Activity log for ongoing organizational activity at the University

These seven areas require different data collection techniques that rely primarily on qualitative methods, including unobtrusive measures, security surveys, interviews, and analysis of written documentation on socio-historical information of HAZMAT policies and practices. According to Maxwell (2005, 93–94):

> Collecting information using a variety of sources and methods is one aspect of what is called *triangulation* (Fielding & Fielding, 1986). This strategy reduces the risk that your conclusions will reflect only the systematic biases or limitations of a specific source or method, and allows you to gain a broader and more secure understanding of the issues you are investigating.

Unobtrusive data was gathered through observational surveys. Interviews were conducted with former and current faculty and staff who were acquainted with the history and status of HAZMAT at the University. Federal HAZMAT regulations and the HAZMAT practices of the University's benchmark institutions were also collected. From this data, recommendations for improved practices were formulated for universities that use biological elements in their research activities.

This study relies on documentation provided by government agencies, documentation of current HAZMAT procedures, other universities' organizational structures and HAZMAT procedures, and historical records outlining the evolution of the University. Obtaining historical HAZMAT documentation from the University proved to be difficult. To establish a timeline of events, historical information on the University was gathered by analyzing past annual reports produced by both the Research Think Tank and the University. Staff members who were identified as potentially having copies of the required documentation were contacted concerning the status of the documentation. Former Think Tank employees were particularly helpful concerning the early days of the University as it transitioned from a conglomeration of research laboratories into an institution of higher education. These individuals not only offered documentation they still possessed but also provided recollections that were very insightful. Federal legislation and federal agency guidelines applicable to the experimentation and use of elements were collected. An example of this type of legislation is the Public Health Security and Bioterrorism Preparedness and Response Act of 2002 and the Homeland Security Chemical Facility Anti-Terrorism Standards 2007 that were, among other things, used to establish federally mandated standards on research centers and university research. A federal investigative report on security with selected agents was also used to illustrate the kinds of guidelines imposed on the universities by government doctrines.

EVOLUTION OF RESEARCH TOPIC AND THEORY

Originally the intention of the research was to implement some policy or practice with an organization. A topic had been selected that would allow for the development of software that could track various types of HAZMAT within the organization. The idea was that the software would allow for better intelligence and more effective crisis response and that the software should also allow organizations to effectively track assets to locations and responsible individuals. A new research topic was then developed that would involve the adoption of the software as part of the organization's policy.

Additionally, the research topic had to be grounded in a theoretical core and a working hypothesis (Glaser and Strauss 1967). The result led to the investigation of various security levels for research activities that an organization could apply without negatively affecting researcher productivity. The working hypothesis, based on the Los Alamos model, was that increases in security levels fostered by external threats or regulations need not result in negative impact to productivity (Rhodes 1986). The hypothesis envisioned various loci for security. On one end of the spectrum was the individual researcher with accepted disciplinary standards for lab protocols (investigator-centered security and control) and at the other end was organizationally dominated security and procedures.

During the summer of 2004, data-gathering activities commenced with an initial security survey of the organization's security practices in research facilities. The research required conducting several security surveys using unobtrusive methods to collect data on how the organization handled security of research activities. Unobtrusive observation allowed the researcher to see how the organization performed research activities with the current security measures in place. The first question that arose was how did HAZMAT security in the research facilities develop into its present state of affairs? To answer this question, the research focused on the evolution of the University's HAZMAT policies, procedures, and practices in a changing regulatory environment. A total of four security surveys were performed on the main research facilities where HAZMAT materials were located. Observations gathered during the security surveys led to developing the following explanations:

1. The current HAZMAT situation developed primarily because of the lack of resources for researchers as the University evolved from a purely research operation into a more broad-spectrum institution. Resources were diverted into new development.
2. As a result of continued growth in other areas, the University has been slow to adjust to the increasing regulatory environment external to the organization.

These statements were generated from the data, collected using the various methodologies. As stated by Glaser and Strauss (1967, 1):

> We believe that the discovery of theory from data—which we will call grounded theory—is a major task confronting sociology today, for, as we shall try to show, such a theory fits empirical situations, and is understandable to sociologists and layman alike. Most important, it works—provides us with relevant predictions, explanations, interpretations and applications.

The statements maintain validity when compared to all other data that were collected from interviews, observations, and archival documents.

ETHICAL CONSIDERATIONS IN RESEARCH

Throughout the research, decisions were made to release selected information to the University's staff in Environmental Health and Safety, the University Police Department, Sponsored Projects, and other administrators at the University. These decisions were based on the determination that there was a real and present danger to life and property, and liability for the University staff, students, and faculty. Furthermore, the researcher is an employee of the University and is bound by ethical obligations that are explicit in the conditions of employment. As Leedy and Ormrod state (2001, 107):

> Researchers should not expose research participants to undue physical or psychological harm. As a general rule, the risk involved in participating in a study should not be appreciably greater than the normal risks of day-to-day living. Participants should not risk losing life or limb, nor should they be subjected to unusual stress, embarrassment, or loss of self-esteem.

During the security survey, I discovered information that revealed key areas that were not in compliance. I reported this to my direct supervisor, who instructed me to report the information to the proper authorities. For example, during one weekend survey, the North Laboratory door was unlocked and the alarm system for that building was not connected to the University Police Department (it was a stand-alone system that had been installed years ago). Another example was the discovery of an unlocked external basement door that permitted access to areas where biological HAZMAT and other laboratory activities took place. Although these doors were locked after this information was given to the University Police Department, other actions and behaviors that I will describe later in this report remained unchanged. Thus, the data that was released to authorities cannot be said to have altered the research site in a significant manner. However, discovering and reporting noncompliance with safety protocols raises important ethical issues about conducting fieldwork in settings where the researcher has explicit obligations to the safety of the campus community.

RESEARCH METHOD: INTERVIEWS

Benefits of Interviews

The purpose of the interviews was threefold: first, to supplement documentation that was gathered; second, to learn where documentation may reside; and third, to ascertain a sense of how policy was enacted in practice. In some cases, documentation did not exist and interviewing was the only method available that could establish when certain events occurred. Interviews were likewise vital in completing the University timeline accurately and thoroughly. Interviews also provided insight on why certain actions were taken or why documentation was constructed in a certain fashion. Interviews were conducted with faculty members of the Research Think Tank, current research faculty, former and current heads of relevant faculty senate councils, and the former head of the faculty senate. Both the interviews and the observed current practices were necessary to develop recommendations to the faculty and administration for improved HAZMAT practices. For information on the history of the University and current HAZMAT practices, in-depth qualitative interviews were conducted with certain faculty members who currently use or had used HAZMAT.

Using open-ended questions while interviewing faculty and staff who were knowledgeable of the University's HAZMAT allowed respondents to provide additional information with their answers. The semistructured nature of the questions also avoided some of the problems Maxwell (2005) discusses:

> Some qualitative researchers believe that, because qualitative research is necessarily inductive and "grounded," any substantial prior structuring of the methods leads to a lack of flexibility to respond to emergent insights, and can create methodological "tunnel vision" in making sense of your data. (p. 80)

All respondents were asked the same questions for a given topic (see Appendix C). The impetus for maintaining the integrity of the interview instrument is evident in Drew and Hardman's (1985) description of how changes in an instrument can negatively affect internal validity:

> In this context, instrumentation is defined as "changes" in the calibration of a measuring instrument or changes in the observers (Campbell and Stanley, 1963, p.175) that may influence the scores or measures obtained ... If something occurred midway through the experiment that changed the calibration of the instrument, such as adjustment by a well-meaning service representative, all the data collected from that point on would be systematically different from those data collected before the change. (pp. 134–135)

For the in-person interview, the questions were carefully ordered to encourage the respondent to provide additional information and to establish a relaxed environment for successful interaction between the interviewer and the respondent. Each interview session lasted approximately an hour. Questions that were deemed controversial were reserved for the end of the interview session (Warren and Karner 2005). Semistructured interviewing was relied upon because the objective was to obtain as many answers from the respondents as possible. As stated by Lofland and Lofland (1995, 18): "Intensive interviewing, also known as "unstructured interviewing," is a guided conversation whose goal is to elicit from the interviewee (usually referred to as the "informant") rich, detailed materials that can be used in qualitative analysis."

In addition to providing details about faculty members' length of employment or whether they worked on HAZMAT policies, the interview process would often reveal names of other staff or faculty members who held information on HAZMAT or information on evolution of the University. This snowball sampling method became a resource for recruiting additional participants. For example, one faculty member who worked on a university committee that recently formulated HAZMAT policies mentioned a colleague who had participated in the development of prior HAZMAT policies. This faculty member was then interviewed to obtain data on the evolution of HAZMAT policies for the University. A number of questions were designed to intentionally mask the type of information being solicited (see Appendices C through F). Often several questions were asked during the interview process to obtain one piece of information. This was done not only to gain information on a particular subject but also to minimize respondent's potentially negative attitude toward the questions.

Limitations of Interviews

Interviews require a researcher to rely on a respondent's memory for data, yet memories of past events can be inaccurate or unclear (Leedy and Ormrod 2001). If interviews are used, they should be employed in conjunction with other methods that can aid the researcher in verification of the information provided. Another drawback to an interview is the problem of ensuring that the observer recorded the information accurately (Drew and Hardman 1985). Although tape recording the interviews would have been preferable, there was a potential disadvantage in using a recording device because the subject matter was controversial and the presence of a recording device might have made the respondent nervous or prone to self-censorship. As a result, it was decided to take extensive notes during the interviews; even so, the recording of the data was not as accurate as it might have been if recording devices had been employed.

Questionnaire versus Face-to-Face Interviews

The decision to utilize face-to-face interviews instead of pencil-and-paper questionnaires was a deliberate attempt to counteract the sensitive and extremely controversial nature of the questions. Moreover, paper questionnaires inevitably limited the topics and were too inflexible for the subject matter. Since HAZMAT policies or procedures have been investigatory in nature, a paper questionnaire may also be considered somewhat antagonistic to the faculty and therefore may impact their willingness to participate (Drew and Hardman 1985). Likewise, a paper questionnaire does not convey tone or friendliness to the respondent, which may limit rapport and consequently affect the participant's willingness to disclose sensitive information. In addition, the respondent can only answer the questions posed to them on the form and cannot elaborate on related issues. Moreover, a paper questionnaire cannot explain the question to the respondent if he or she is unclear as to what type of data the question is attempting to retrieve. As described by Drew and Hardman (1985, 198):

> There are many intricacies to questionnaire studies that are not apparent on the surface, and some of them relate directly to the instrument … It must, in large part, stand on its own because a researcher is not usually present to prompt a response or clarify areas in which the subject may be confused.

A paper questionnaire could potentially cause respondents to become defensive, which would discourage them from completing and submitting the form. The fixed format of a questionnaire does not permit the respondent to provide additional data that is sometimes spontaneously volunteered in a face-to-face interview.

Data Collection

Five different interview schedules were generated to investigate past or present issues with HAZMAT. Previous interaction with respondents was very limited. One respondent was known before the start of the research and had interactions with various staff members as a consequence of having a common employer. It is not believed that the employment or previous contact with the respondents biased answers during the interview sessions since this topic was not previously discussed with the respondents. A perception exists that having a common employer did improve access to faculty and staff for the project. As the principal investigator, all interviews were administered personally and each interview schedule is detailed next.

The first interview schedule (Appendix B) dealt with the evolution of HAZMAT at the University. There were nineteen questions posed to six respondents. Each interview lasted from an hour to an hour and a half. The interview schedule was printed before each interview and the respondent's answers were handwritten. No tape recorder was used since possible resistance to answering questions was anticipated with the use of a recording device. All interviews were conducted at the faculty member's office or home, so that the environment conveyed a sense of comfort for the respondent. As previously mentioned, respondents were selected due to their length of service at the University and the nature of the scientific research the respondents conducted. The questions were formulated to obtain information on how conditions evolved and to determine if any centralized HAZMAT policies previously existed.

The second interview schedule (Appendix C) contained questions for current staff members of the University on current practices and procedures for HAZMAT. There were twelve questions on the instrument that were specific to the staff members' knowledge on current practices and procedures for HAZMAT. Two staff members were interviewed for a period of an hour to an hour and a half in their offices. Again, the interview schedule was printed before each interview and answers by the respondents were handwritten. Unlike the questions to faculty members, the staff members who were interviewed were not directly responsible for HAZMAT and as such were asked more directly about current practices and procedures according to the University's policies. Both staff members interviewed had contact with the faculty members performing research and also had knowledge about current ongoing scientific research federal regulations pertaining to the University. The staff members were asked questions about how the University was complying with federal guidelines and regulations, in other words how policy translated into practice.

The third interview schedule (Appendix D) was used to obtain information for the University of Texas Southwestern Medical Center's (UT Southwestern) practices and procedures for HAZMAT. One respondent was interviewed for a period of an hour to an hour and a half. Fifteen questions were included in the interview schedule. The respondent was interviewed over lunch in an informal and neutral environment. The questions posed to the respondent were used to gain information on the respondent's knowledge of the current practices and procedures used at UT Southwestern as well as which practices and procedures caused the organization difficulty.

The fourth interview schedule (Appendix E) posed questions to faculty members on the evolution of the HAZMAT policies and the status of current HAZMAT policies. Three respondents were interviewed for an hour to an hour and a half. The purpose of this interview was to gather more data on the HAZMAT policies and procedures. The interview instrument was composed of twenty questions, which investigated the current state of HAZMAT, the status of the HAZMAT policies, and the evolution of HAZMAT policies and procedures. This information was not obtained in prior interviews and required additional data. The respondents were interviewed in their offices.

The fifth and final interview schedule (Appendix F) was used to fill in voids of information in the history of the University that previous interviews had been unable to obtain. One respondent was used to gather data on the history of the University. The respondent was a former employee who had occupied a high-level position in the administration at the Research Think Tank. The interview lasted about an hour and the interview instrument had fourteen questions. The respondent was interviewed at the University in a departmental office, since the respondent was no longer employed full-time at the University. The researcher has actually known the respondent for two years but even though they had had numerous interactions, the focus of this research was never a topic of discussion. Since the respondent was a previous employee, questions were asked the respondent in a more direct fashion (i.e., how did the refrigerators get into the hallway?). Again, no tape recorder was used since potential resistance to answer questions was anticipated to increase with use of that recording device.

RESEARCH METHOD: UNOBTRUSIVE OBSERVATIONS

Rationale for Unobtrusive Observations

The researcher decided that unobtrusive observation in the form of a series of observational security surveys of actual HAZMAT campus practices would be an effective method for gathering data. As described by Webb et al. (1973, 50):

> The outstanding advantage of physical evidence data is inconspicuousness. The stuff of analysis is the material which is generated without the producer's knowledge of its use by the investigators. Just as with secondary records, one circumvents the problems of awareness of measurement, role selection, interviewer effects, and the bias that comes from the measurement itself taking on the role of a change agent.

A descriptive analysis was then produced establishing current HAZMAT practices at the University. For each building a checklist was used to see if certain elements were present (e.g., biological containers present in the hallways) or specific techniques were employed (e.g., are laboratory doors locked). Each security survey shared a common checklist of items that were observed throughout each research facility that used biological specimens.

Data Collection for Unobtrusive Observation

To view unobserved behaviors, the researcher decided to assimilate with other persons that might be encountered in the environment where he was conducting the survey (Webb et al. 1973). To blend in with any students the researcher might encounter, he dressed in a T-shirt, ball cap, and either shorts or blue jeans. The researcher entered research areas that were unlocked and accessible to anyone. On two security surveys, the researcher was accompanied by the crime prevention coordinator to verify the researcher's observations for future corrective actions by the University police and to explain the researcher's presence in certain research areas if they were stopped and questioned by the University Police Department. In an attempt to assimilate into the general campus population, the crime prevention coordinator did not wear clothing that identified him as working for the University police and carried only his department-issued radio. There were four security surveys performed during the data-gathering phase to verify consistent behaviors or actions. As stated by Drew and Hardman (1985, 202):

> Observation studies may also use the interval method of recording data. Such procedures focus on whether or not a particular behavior occurred during a given interval. Basically, the observer would simply place a check mark or a code symbol on the recording form if the behavior occurred during the specified time interval.

The researcher intentionally performed the security surveys during times when the research facilities should have been most secure—during the weekend when no classes convened and when minimal faculty are present.

Limitations of Unobtrusive Observations

Unobtrusive observations have numerous weaknesses and limitations. The disadvantage of using observational techniques is that it requires multiple points of data to gain an accurate depiction of behaviors or actions. Moreover, observation does not address the fact that behaviors or actions may be different during a particular time due to a certain event (e.g., when class is in session), that behaviors could be disguised or hidden during observation, or during the presence or absence of a particular individual (e.g., head of a department). Indeed, the interval observational method may not capture this type of anomaly in behaviors or actions.

Another problem that arises with observational techniques is the ability of the researcher to gain access to certain areas unobserved. Due to the nature of the subject matter, at times it was not deemed safe to enter certain facilities to make observations (e.g., to open the door to a radioactive room). For other observations of laboratories, the researcher would not have been able to be present without the faculty members, students, or staff noticing or potentially allowing the researcher to observe actions that occurred inside the laboratory.

RESEARCH METHOD: ARCHIVAL DOCUMENTATION

Use of Archival Documentation

The initial research consisted of obtaining written documentation and information on existing HAZMAT policies, practices, and procedures at the University. As stated by McNabb (2002, 89), "To achieve qualitative study objectives, researchers analyze the interaction of people with problems or issues. These interactions are studied in their context and then subjectively explained by the researcher."

Acquiring this documentation first required a discovery process to verify if any documentation existed, and if not, investigating where the documentation was once physically located. As described by Webb et al. (1973), archival data serves several purposes.

> Besides the low cost of acquiring a massive amount of pertinent data, one common advantage of archival material is its nonreactivity. Although there may be substantial errors in the material, it is not usual to find masking or

> sensitivity because the producer of the data knows he is being studied by some social scientist. This gain by itself makes the use of archives attractive if one wants to compensate for the reactivity which riddles the interview and the questionnaire. (p. 53)

Due to the age of the University, many of the personnel who initially started the organization as the Research Think Tank were unavailable, and therefore archival documentation was the only source of data that could be gathered in certain cases. Other institutions' HAZMAT practices and procedures documentation was researched to determine gaps in the existing University's practices for HAZMAT security and safety. Furthermore, information on federal guidelines and legislation was collected through either federal agency documentation (e.g., CDC/NIH biosafety guidelines for laboratories) or United States Congressional Acts (e.g., Public Health Security and Bioterrorism Preparedness and Response Act of 2002).

Limitations of Archival Documentation

The primary disadvantage of using archival documentation is determining whether the documents in question actually contain information that is valid or pertinent to the situation at hand (Leedy and Ormrod 2001). Another potential challenge is the proper interpretation of the original intent of the documents (Leedy and Ormrod 2001). Both problems in the project were addressed by supplementing the documentation with interviews and observations.

RESEARCH METHOD: PARTICIPANT OBSERVATION

Log of Participant Observation

During the security surveys, the researcher maintained a log of various HAZMAT situations that could be improved. Working with other departments allowed the researcher to gather information that otherwise would have been difficult to obtain. Denzin (1978) notes:

> The participant observer learns to employ multiple methods. Documents are collected and analyzed, interviews are conducted; informants are sought out for their unique perspectives; and direct participation in the

> group's activities is appropriately conceived, participant observation represents the simultaneous use of many methods. It becomes a triangulated methodology and is a direct extension of the naturalistic method. (p. 255)

Cooperating with the Environmental Health and Safety department and the University Police Department on different projects allowed for data collection on departmental operations and responsibilities for HAZMAT. As stated by Webb et al. (1973, 115):

> The control effect is present when the measurement process itself becomes an agent working for change: "the difficulty with control effect in participant observation, and in many other research designs, is that it is unsystematic…" (p. 71) … Dalton (1964) gives an excellent pro and con analysis of participant observation in his commentary on the methods used in *Men Who Manage*. Dalton's pro list is longer than the con one, and he employs the intriguing terminology of "established circulator" and "peripheral formalist."

An example of an agent working for change occurred when the security survey results were given to the chief of police, and the chief requested recommendations to improve observed situations. These discussions were documented as well as the implementation of some of those recommendations. Discussions were held with a number of other departments with regard to operational procedures with HAZMAT and the assistance that the researcher provided those departments in implementing improvements to their operations. The researcher's participation was kept confidential throughout the process. The next four chapters contain data collected from the research methods that were described in Chapter 4.

Disadvantages of Participant Observation

The disadvantages of participant observation involve potential bias that can affect the resulting data. Bias caused by emotional ties to the organization could sway the researcher to act or record observations in a particular manner that would be detrimental to the research (Leedy and Ormrod 2001). To ground the research, other methodologies were used to corroborate findings and verify the researcher's actions during the research project. As with the interviews, archival documentation, and unobtrusive observations, participant observation is used in conjunction with other methods instead of as an isolated methodology.

ETHICAL DILEMMAS OF RESEARCH

Upon completion of my research, which included four security surveys, I reported the findings to my dissertation chair. The original dissertation project called for multiple security surveys utilizing an unobserved method. However, the shortcomings and hazards I uncovered were determined to be too dramatic not to report these findings to the chief of police. On June 8, 2004, the findings were turned over to the chief of police and the crime prevention coordinator. My dissertation chair then advised me to focus on observing the organizational response to the HAZMAT concerns rather than researching faculty productivity against security concerns.

The release of my findings on the University's security shortcomings set into motion a series of events that impacted how the University managed HAZMAT. The chief of police requested a HAZMAT inventory list from the manager of Environmental Health and Safety. The crime prevention coordinator discussed some of the findings with the manager of Environmental Health and Safety especially with regard to the accessibility issues of the North Laboratory. The crime prevention coordinator and the manager of Environmental Health and Safety agreed to lock up the North Laboratory. The crime prevention coordinator reported external doors that were not lockable to the physical plant for repairs, but as of June 14, 2004, no action had been taken by the physical plant on that issue. By releasing the information, organizational actions in some noted shortfalls were altered.

6

How HAZMAT Conditions in the Science Buildings Evolved

HISTORICAL CONTEXT

As with any organization, events cause change. How and why situations develop over time is just as important as how and why an organization was initially founded. Some of the decisions on hazardous materials (HAZMAT) were formulated in the absence of any federal laws or mandates. For example, the Founders building was constructed in 1962, before the joint Centers for Disease Control and Prevention (CDC)/National Institutes of Health (NIH) standards for biosafety in laboratories existed. In 2005 the building was still used by researchers for biological experiments even though, according to my respondents, the facility has not been upgraded to CDC/NIH biosafety laboratory standards. Researcher No. 3 stated that the building was already considered obsolete as a science building in the 1980s when he was originally hired.

RESEARCH THINK TANK

In 1959 the founders of a major corporation recognized that the area in which their corporate headquarters was located could not rely on extractive and other traditional industries and remain economically competitive on a national and international scale. The future economic growth of the region, they reasoned, would require a new foundation in science and technology (Jordan 2002, 6). They wrote that "the region must grow academically; it must provide the intellectual atmosphere which will allow it to compete in the new industries dependent on highly trained and

PICTURE 6.1
1960s aerial photo of D1 and D3 when first constructed.

creative minds" (Graduate Research Center of the Southwest 1961, 10). Referencing the emerging technologically intensive centers of Route 128 around Boston and areas around San Francisco, the founders pointed out that these emerging complexes were gathered around major universities (Jordan 2002, 6). The founders in concert with other area philanthropists created a foundation aimed at creating a research think tank.

The foundation purchased a 968-acre site in 1964 that was basically farmland and built what was for the time a sophisticated scientific research building (Picture 6.1).

The Research Think Tank, over the next nine years, employed an excellent research faculty educated in geophysics, mathematics, biology, and earth and planetary sciences (Jordan 2002, 7). Essentially, the work of these researchers became the core of the emerging university. But unlike national laboratories, the Research Think Tank was initially operated and founded by private citizens instead of the government. According to one knowledgeable informant who worked at the Research Think Tank, it was common for the founders to discuss with the chief financial officer how many shares of stock the founders needed to sell to finance the operation of the Research Think Tank.

When the Research Think Tank was originally constructed, residential housing was virtually nonexistent, roadway infrastructure was minimal, and miles of farmland (in wheat and cotton) surrounded the institution. Essentially the Research Think Tank was established in a wide-open area surrounded by miles of vacant land. (Picture 6.1). The history, location, and arrangement of the Research Think Tank were similar to those of

PICTURE 6.2
Distance picture of D1 and D3 when first constructed.

many government-sponsored labs. According to Researcher No. 6, some scientists were moved from the private university located downtown to the Research Think Tank facility (D1) before the facility was even completed. After completion of D1, the scientists for the Research Think Tank were all relocated at the new facility. D1 still houses biology, classrooms, computer laboratories, and the office of undergraduate studies (Picture 6.2).

Some administrative functions for the Research Think Tank were located in an off-site, leased office building. A cosmic observatory, an annex (D3), and a connecting skywalk to D1 were also completed in 1964. Shortly after D1 and D3 were completed, the Research Think Tank became independent and operated without the private university affiliation.

Since no central oversight existed for HAZMAT policies or procedures, research scientists began to develop a culture of operating independently for their own research activities. Individually scientists may have been interested in what projects other researchers were developing, but they had no incentive to dictate to another researcher how the project should be performed. The focus for the researchers was to perform groundbreaking work in their respective fields of study and the researchers had confidence in the ability of their colleagues to safely carry out experiments. Staff Member No. 1 was initially told upon his hiring that the founders gave the president of the Research Think Tank and the research "giants" unlimited financial support. Staff Member No. 1 described that a biological researcher and his staff were imported into the Research Think Tank from Germany. The researcher brought with him a policy and procedure manual (in German) that he fully intended on using for his research activities. According to Staff Member No. 1, researchers were expected to handle their own biological HAZMAT since the administration did not address HAZMAT issues during that time period.

In 1965, a private association began offering graduate and postdoctoral education at the Research Think Tank. One of the founders of the Research Think Tank envisioned that a university would be established to keep highly talented students from leaving the area. Staff Member No. 1 stated that the founders' intent was to bring the state into the modern era with an emphasis on modern technology and have less reliance on farming and natural resources. It became apparent to the founders that additional sponsorship would be needed if the institution were to add the educational dimension the founders had first envisioned, and consequently, in the mid-1960s efforts were begun to move the think tank into a state university system. In 1969, the University officially came into existence with the Research Think Tank buildings and with restrictions. The establishing legislation restricted the University to a graduate education mission until 1975, at which time it could enroll upper division students but not freshmen and sophomores. The University could offer only a restricted menu of graduate degrees. In 1970, the enrollment was 45 graduate students (Jordan 2002, 8).

The addition of educational services to the Research Think Tank began to change the purpose of D1; however, raising funds for research activities was still deemed more important. In 1965 the expansion of educational programs was halted until new research facilities could be constructed. The F1 facility was constructed at this time. Currently this building houses HAZMAT waste as well as several physics laboratories.

While new facilities were constructed and D1 expanded its mandate to include education as well as research, scientists were conducting research in the fields of genetics, material science, space sciences, geosciences, and physics. One experiment in particular dealt with radiation damage to biological specimens and the capability of simple biological organisms to protect or repair themselves after sustaining such damage. A facility was installed to contain a device that generated intense magnetic fields. The

Research Think Tank was also one of the first institutes to receive the new IBM 360 mainframe.

The original campus master plan in 1964 allowed for additional research facilities, not the expansion of educational facilities. In 1965 a land use planning committee recommended that the Research Think Tank set aside land for a research park that would include industrial research laboratories. In 1966, the I1 facility was added to the Research Think Tank and scientists specializing in mathematics and physics occupied the building. I1's intended use was as a research facility. This time period also had very little state or federal regulations for performing research activities with regard to biological HAZMAT.

EVOLUTION FROM RESEARCH THINK TANK TO GRADUATE UNIVERSITY

In 1969 the Research Think Tank became part of a state university system and began to admit master's and doctoral students in the physical sciences and computer science. This permitted the continued emphasis on research while providing educational resources for local residents.

The arrival of even a small number of students created changes in the use of physical space. D1 had not been constructed with classrooms in mind. Soon after 1969, refrigerators in D1 used to house specimens for biological experiments were moved into the utility corridors to accommodate classrooms. After an inspection of D1, a fire marshal, citing safety concerns, issued a directive to remove the refrigerators from the utility corridors. This forced the University to "temporarily" move the refrigerators to the main hallways since no other space was available; the refrigerators remain in their temporary location forty-three years later. E1, which housed chemistry and biology laboratories, was completed in 1973. Federal and state regulations for HAZMAT gradually increased during the time period of 1969 to 1975.

Throughout the transition from Research Think Tank to university, the researcher-centered culture of the Research Think Tank predominated. The faculty of the Research Think Tank, now nascent university, were world renowned and logically expected high standards of work from one another but afforded to one another the individual responsibility to establish protocols for their own labs. The "central administration" of the

Research Think Tank from the beginning was composed of scientists who espoused a similar ethos. In 1971, the new university had its first nonscientist as president but no change occurred, nor was recognized as needing to occur, with regard to HAZMAT policies or procedures. The new president was charged with building new facilities aimed at the arrival of upper division students among other matters (e.g., hiring over 100 new faculty). Thus an informal "safety" culture prevailed throughout the transition from Research Think Tank to university.

TRANSITION TO AN UNDERGRADUATE AND GRADUATE UNIVERSITY

Beginning in 1975, the University admitted its first upper division undergraduate students. Several new academic buildings had been constructed primarily to provide teaching facilities and office space for nonscience faculty. However, there was no corresponding expansion of research facilities during this time period, which left faculty to perform research in facilities that were cramped and in some instances obsolete. Research faculty now included teaching as an activity. The University offered degrees in other academic areas such as liberal arts and business to a growing body of commuter students. New residential areas, roads, and other aspects of infrastructure began to appear around the university. This ended the institution's isolation and limited the kind of research that it could safely perform within the increasingly populated area.

The following decade represented a transitional phase for the University. In 1979, the Institutional Review Board (IRB) established guidelines for research involving human subjects. The IRB policies were the first known attempt to establish centralized controls over university research activities and even then, according to one committee member, the discussions centered almost exclusively on research involving human subjects. A Biosafety Committee was also established in 1979 to which the president of the University appointed committee members. There were no annual reports (archival documentation) found to account for any activities of the Biosafety Committee. Initial duties of the Biosafety Committee are unknown since the original policy memo could not be found. Research laboratories focused on other materials, from all apparent data, and continued to operate under protocols established by individual researchers.

According to respondents, students were rarely found in the main corridors of the research facilities and instead were more commonly seen in C1, J1, D2, and A1 facilities where the majority of classrooms were located. The completion of C1 in 1988 capped a decade of construction and completed the University's transition from research to instruction. C1 initially contained classrooms, some laboratories, and administrative offices. According to a respondent, the Institute for Environmental Studies was located in C1 until the institute was dissolved in the mid-1980s. The Engineering and Computer Science dean's office suite was initially located on the third floor of C1. One wing of C1 was progressively transformed to house biology and Behavioral and Brain Sciences laboratories. Recently however, C1 has become primarily a building for the administration of the University. According to respondents, students were not seen in the corridors or D1 or E1 since few classes were taught to students in those two buildings, and the building for the Engineering and Computer Science program, B1, did not yet exist. During this time, more federal and state regulations were enacted with additional Environmental Protection Agency (EPA) and NIH guidelines for HAZMAT. The joint CDC/NIH biosafety guidelines (recommended practices, not required) were established in the 1980s as recommended practices for laboratories contending with biological elements.

UNIVERSITY ADDS LOWER DIVISION UNDERGRADUATES

In the early 1990s, the University admitted lower division undergraduate students. Their presence necessitated new services and facilities such as student housing in the form of apartments located on campus grounds. The first segment of apartments was completed in 1989 for graduate students. Apartments would eventually expand to house undergraduate students as well until early 2010, when the University began constructing dorms. Students were still primarily commuters attending evening courses even after the first apartments were constructed. With an increase in students enrolled in lower-level science courses, more space was required for teaching laboratories and classrooms. Satellite facilities were constructed to separate sponsored research from laboratories devoted to student instruction. In 1991, G1 was completed. Laboratory animals were then removed from D1 and housed in G1. In 1992, the university constructed a space for a linear particle accelerator. The equipment was installed in a facility away from student and staff main traffic areas. The refrigerators in D1 hallways remain where they had been for the past decade; there was still no place to secure them (as of 2009). The addition of B1 and a connecting skywalk between B1 and E1 increased foot traffic through the research facilities. Before B1 was constructed, students rarely traversed the campus through the main corridors of the research facilities.

As the new millennium dawned, the search for more research space continued at the University. E1 was renovated in 2001 to provide research space for a sickle cell research center and a nanotechnology institute. The University also constructed H1 for research (in 2006), K1 for research and instruction (in 2010), acquired M1 for research and instruction (in 2002), acquired N1 for research and instruction (in 2005), constructed B2 for research and instruction (in 2002), acquired O1 for administration and research (in 2010), and acquired Q1 for research (in 2006). Additionally renovation began for D1 and D2 in 2007, which would convert some areas from research to instruction. C1, which initially contained only classrooms, some laboratories, and administrative offices, had been progressively transformed to house biomedical laboratories. With the transformation of C1 into research and classroom space and the advent of skywalks to B1, an increase of student, staff, and faculty traffic occurred not only through the newly renovated areas but the older research areas as well. The original campus design did not take into account the need to accommodate an increase in traffic throughout the research facilities.

The new facilities that are currently being constructed are stand-alone facilities intended to solve the problems with D1 as a research facility. Four new buildings have been built away from residential areas and separate from the main part of the campus. By building the new facilities apart from other campus buildings, the isolation should reduce the amount of traffic of personnel and students who do not need access to those facilities. Furthermore, the items used in research activities will be in a more secure location (e.g., refrigerators).

The University's proximity to residential areas and the increased enrollment of undergraduate students had renewed safety concerns. The University Safety Council was established in 1992 to develop policies to improve safety for university employees and students. In 1994 the first University Safety Manual was compiled. Due to National Science Foundation requirements for research projects that experimented with recombinant DNA, an Interim Biological Safety Manual Section was compiled in 2003 and distributed June 13, 2005. The Safety Manual was revised in 2003 and was also distributed June 13, 2005. In 2004, The University Police Department was evaluating a new emergency operations plan (EOP) for the university which has now been distributed (2012). In July 2005, the Biosafety Committee Charge, the Safety Manual, and the Biological Safety Manual were finally posted online at the University. However, annual reports that were required by the Biosafety Committee had still not been located as of 2005. According to one respondent, such reports are not known to exist.

The Institutional Animal Care and Use Committee (IACUC) updated animal care policies and procedures in 2005, but the date of origin of the policies is unknown. From 1990 to the present a host of federal and state regulations have been instituted for HAZMAT used for research purposes (e.g., the Patriot Act of 2001).

OBTAINING BACKGROUND INFORMATION

Faculty members who were employed at the University in the mid-1960s to the 1980s in a variety of research fields and who would had been familiar with the organizational decision-making process and the condition of D1 during that time period were interviewed. These faculty members were asked a series of questions that are listed in Appendix C.

WHO MADE DECISIONS WITH REGARD TO HAZMAT?

The researcher asked all of the respondents the question: "At which level were decisions (regarding research) made?" All respondents stated that individuals involved in research activities made their own decisions with regard to safety and security issues in the absence of administrative oversight. Researcher No. 1 commented that faculty enjoyed working at the University because the administration "left you alone to perform your research." The general consensus from the respondents was that although the administration did not assert itself in the daily operation of laboratories, it did steer faculty members to perform research activities that could be supported by the institution's budget. This was in contrast to how research was supported in the University's prior incarnation as the Research Think Tank. Researcher No. 6 stated that Research Think Tank administration often provided a great deal of financial support until research scientists could acquire sponsored contracts from outside agencies such as the National Aeronautics and Space Administration (NASA) or the National Science Foundation.

HOW THE UNIVERSITY HAS EVOLVED SINCE THE EARLY 1970S

The respondents all believed that the influx of students, especially undergraduates, has changed the manner in which they perform their jobs. Researcher No. 3 said there was more bureaucracy between the researcher and the upper administration. Enrollment increases forced a reduction in research space to accommodate more classrooms. Two faculty members stated that due to an increase in government regulations and the evolving nature of research conducted at the University, D1 is now obsolete as a science building.

HOW D1 BECAME A HAZMAT ISSUE

The primary question the researcher wished to answer was how D1 became such a HAZMAT issue. D1 was the first building constructed at the Research Think Tank facility. Completed in 1964, D1 was intended only for research and did not include classroom facilities. Initially this structure was considered a highly advanced research facility. When the Research Think Tank became the University and graduate enrollment increased, classroom space was derived from existing research space. Faculty relocated their refrigerators to the interior utility corridors. A fire marshal's report concluded that locating refrigerators in the utility corridors posed a hazard. These refrigerators were soon relocated to the main corridors, where they remained as of 2012. With the construction of new research facilities, however, the University has begun to progressively move these refrigerators out of the main corridor areas and into the new research facilities. As these refrigerators are removed from the corridors, the University has begun to renovate these areas into different usable space dependent upon the requirements needed for instruction and research. The researchers initially surmised that the main corridors in D1 were wide enough to accommodate both refrigerators and foot traffic. Researcher No. 6 revealed that a basement was excavated in D1 after its construction to offset the shortage in adequate research space. Researcher No. 6 also stated that laboratory animals were housed on the fourth floor of D1 for a significant period of time. The animal care facility was eventually removed from that floor to G1. G1 in turned has since been demolished and those facilities for animals were moved to the new research facility of H1 in 2006.

Questions were asked about the radioactive material depot on the fourth floor of D1. Researcher No. 3 stated that the department stored radioactive material there "because no one ever goes up there." Questions were also posed about the inventory control of biological specimens. The respondents answered that no such inventory control was maintained. One researcher stated that his laboratory currently writes its chemical inventory on index cards so that the information can be readily relayed to the administration if and when such information is requested. Researcher No. 3 objected to the "unreasonableness" of what was required for inventory control such as the amount of water the laboratory had available. Researcher No. 3 added: "The few controlled substances were in such small amounts that they were not going anywhere." Four of the researchers concluded that the only solution to the biological HAZMAT situation in D1 was completion of the new natural science and engineering research building so that D1 could be decommissioned as a science building. Another issue that could cause a potential problem was the lack of a centralized laboratory for equipment that was shared by the entire biology department. Without such a facility, equipment was housed in individual offices or laboratories, necessitating the trafficking of specimens through common open areas and inviting a greater potential for contamination.

ORGANIZATIONAL RESPONSES TO KNOWN HAZMAT ISSUES

Some actions that have been observed refer to daily operational aspects of the University with students, faculty, or staff. Other actions have been recorded as I have acted as a participant with the Environmental Health and Safety (EHS) Department, the University Police Department, and a variety of departments working to improve conditions at the University. Although a few of the observed actions are somewhat negative, the University has for the most part improved in most areas with HAZMAT and should be commended for progressively addressing difficult issues.

Several events occurred after permission had been granted to investigate security at research universities in relation to research productivity. Additional program applications were developed for the existing Logistical Tracking System (LTS©)* to enable university personnel to track HAZMAT. The LTS application had initially been developed by me and three other employees listed in the Acknowledgments section. I then demonstrated the application to both the University Police Department and the Environmental Health and Safety Department. The University's Environmental Health and Safety Department specified certain modifications to improve the application's functionality.

I met with personnel from the Environmental Health and Safety Department on June 28, 2004, to assess their needs and concerns. The personnel stated that their department was understaffed by at least two members. They currently had three employees in their department; most institutions of similar size to the University had at least seven employees devoted to HAZMAT issues. The personnel stated that they have had difficulty obtaining lists of chemicals and biospecimens from research professors and are forced to conduct on-site inspections of research laboratories to ensure safe conditions. I informed them that LTS could capture that kind of information. Although this department assured me they would use LTS Wednesday or Thursday of that week, this action did not occur. The effort to capture this information would continue to be hampered as of 2012 due to turnover in Environmental Health and Safety personnel and not having a champion who wanted to input the information into an electronic system.

On October 11, 2004, I conducted a facility survey at the University's remote campus. I observed that the staff and faculty periodically confronted me regarding my purpose at the facility and my identity. Since the remote campus had preschool programs at that campus, the staff and faculty were more alert to outsiders appearing at the facilities. The

* LTS has a copyright that is held by the University, Nicolas Valcik, Danald Lee, Dr. Patricia Huesca-Dorantes, and Tarang Sethia. Eleven research assistants provided additional programming support for LTS, including Rajesh Ahuja, Mohit Nagrath, and Shalu Agrawal.

remote campus personnel's response to unknown visitors was completely opposite to that of the staff and faculty at the main campus, who did not appear to be concerned about an unknown person walking through their research facilities.

On February 1, 2005, I was directed by the dissertation committee to narrow the focus of my research to biological HAZMAT. However, I kept all of my previous research on chemical, radiological, and hazardous waste observations and information to potentially use at a later date. The following day, on February 2, 2005, I witnessed a delivery of live amphibians that were contained in a cardboard box perforated with air holes to an administrative assistant's desk that was located in A1. I discovered that the university had disposal regulations and policies in place for animals but did not have a policy in place for delivery of animals to university grounds. On February 4, 2005, I learned that a laboratory that processed tissue samples was inadvertently pumping carcinogenic fumes into the building's ventilation system, which fed directly into an administrative staff member's office in C1. This occurred for over two and a half months because the laboratory's ventilation hood system had broken and was not fixed. I verified the particulars of this incident with two other sources.

On February 8, 2005, I witnessed a student transporting a biological experiment in an open rectangular dish from one laboratory to another in C1. This student was protected only by a pair of latex gloves. The student not only walked through a carpeted public corridor but also passed several staff members who were walking to their offices. This was the same day I was informed by an administrator that the University did not abide by CDC/NIH guidelines for laboratories using biological HAZMAT.

Based on this information I decided to interview an administrator who will be referred to as Staff Member No 2. I discovered through the interview that principal investigators were responsible for ensuring that "restricted persons" (as defined in the Patriot Act of 2001) did not have access to the principals' research. As of March 4, 2005, there was no mechanism in place to determine if a student, staff, or faculty member is considered a restricted person. As of March of 2005, there was no written policy to determine who has access to the research facility for animals or the transportation of animals across campus. G1, which had housed animals for experimentation, undergoes a surprise annual inspection by the U.S. Department of Agriculture (USDA) and had passed all inspections; however, the facility was not certified for a particular biosafety level (as of 2006). Staff Member No. 1 also stated that physical plant and the Office of Environmental Health and Safety were responsible for specification of laboratory construction. I then obtained a copy of the new HAZMAT policies and procedures that had still not been approved or distributed as of March 4, 2005.

On March 9, 2005, I observed a student pushing an open cart of beakers and test tubes containing biological material from C1 to D1 through the skywalks. The student struggled to keep the containers from bouncing off the cart as the cart rolled over rough flooring and doorjambs. This incident occurred over spring break so there were no other students present in the hallways.

On July 13, 2004, I met with the police chief and the crime prevention coordinator to discuss security issues and protocols for the clean room in B1. We reviewed a videotape taken on July 8 by the crime prevention coordinator that clearly showed that the clean room's emergency exit doors were propped open. These doors had been previously pried open from the outside, as evidenced by missing rubber between the doors near the lock mechanism. To further illustrate the lack of existing safety protocols at the University we discussed an incident that took place on July 9, 2004, in which a laser beam had accidentally hit a person in the eye. According to Staff Member No. 4, no one was seriously injured in the incident. On August 10, 2004, the University's Environmental Health and Safety Department shut down all lasers located in the clean room because the clean room did not have proper shielding over glass walls to adequately contain laser beams. It was reported by Staff Member No. 4 that laser beams were penetrating several sheets of glass and were visible on a research building located across from B1.

On July 22, 2004, B1 and B2 were evacuated due an accident caused by an employee who, according to Staff Member No. 5, was attempting to neutralize a chemical but introduced too much of a chemical agent into a mixture, resulting in a chemical fog that filled the building. According to Staff Member No. 7, the Halon system (which uses fine dust instead of water to extinguish fires) was not deployed in the building. One person was taken to the hospital and later released. There were no documents available to inform the police department what chemical caused the fog, making it extremely difficult for the police to protect the public in this situation Staff Member No. 8.

In November 2004, three engineering students were taken to a local hospital after they were exposed to a chemical that was incorrectly labeled and stored in the wrong location. According to Staff Member No. 4, it was presumed that the students were exposed to an acid that could dissolve calcium, including the calcium contained in human bone. After examination at the hospital the three students were released since the chemical did not come in contact with their skin.

In December 2005, after a safety and procedural audit by a review committee was conducted, I interviewed Staff Member No. 9, who had responsibilities within certain research facilities at the University. According to Staff Member No. 9, the audit revealed a potential safety concern from improper storage of chlorine gas. If the containers were damaged and leaked, chlorine gas could be released into the ventilation system, which could lead to massive chemical exposure at one research facility.

On April 22, 2005, I met with Staff Member No. 3 to gather IACUC policies and any policies available for G1. I was asked by Staff Member No. 3 for recommendations that could improve those two policy areas. I recommended that sections be added to IACUC for the transportation and delivery of animals, and for addressing access of restricted persons to research facilities. I also recommended that the online policies be password restricted so that only researchers who needed to reference those documents could access them. Finally, I recommended that cameras be installed inside G1. On April 26, 2005, I introduced Staff Member No. 3 to the crime prevention coordinator so that the issue of cameras for Research Facility One could be discussed. At that meeting the discussion led to the possibility of placing four Web-based cameras with individual IP addresses so that personnel could access the live feed through LTS.

I was asked to give a presentation of LTS on April 29, 2005, to the new Environmental Health and Safety (EHS) director, EHS staff, engineering and computer science personnel, Central Stores, and the crime prevention coordinator. It was agreed that LTS needed to be modified to include a real-time shipping and audit trail for HAZMAT and safety equipment (Figure 6.1).

EHS is currently using the software "EH and S Assistant" to track radioactive elements and waste; however, this software application is apparently not used by other institutions to track chemical or biological HAZMAT. EHS requested that our department modify LTS to track radioactive elements to enable their department to keep an active inventory relative to the floor plans. As a result of this meeting, I learned a great deal about how HAZMAT is transferred through the University. A meeting with Shipping and Receiving personnel determined that a barcode system that assigned

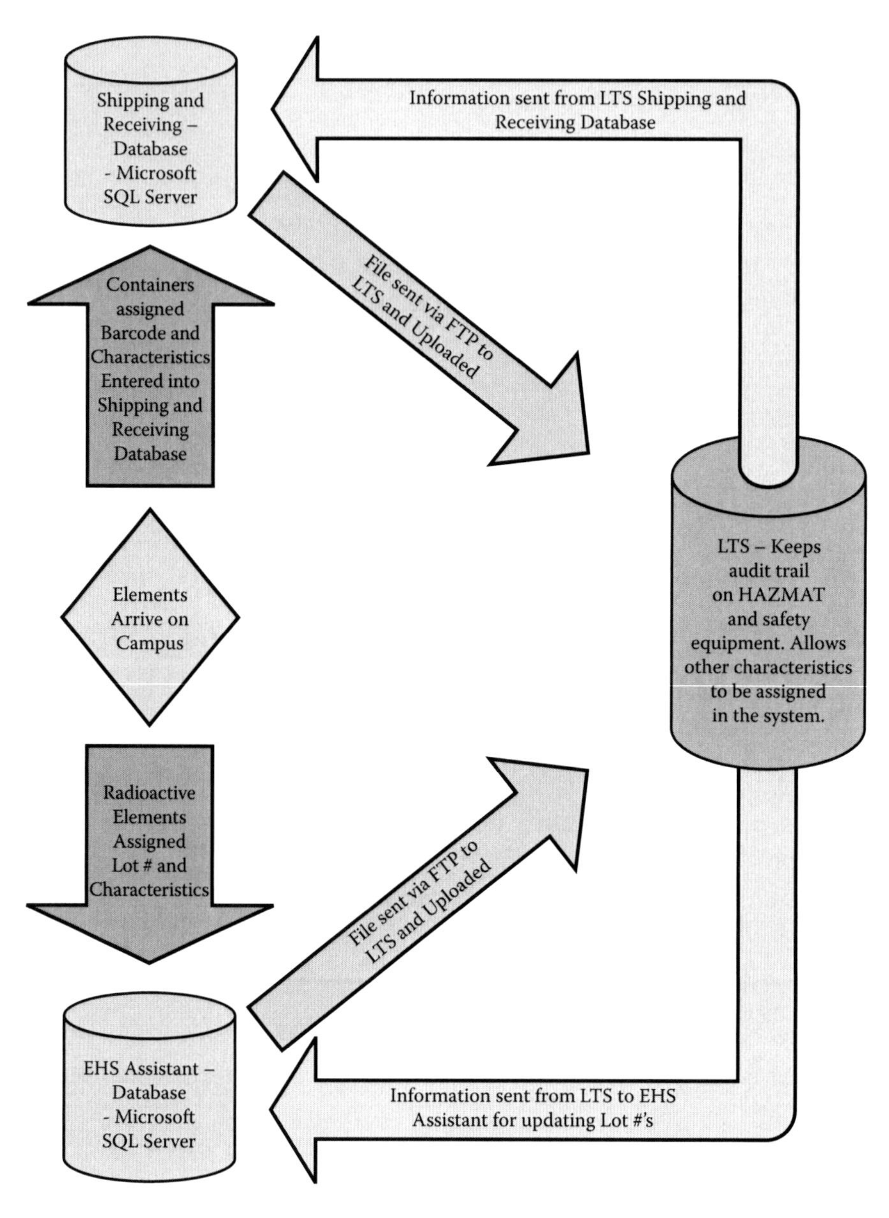

FIGURE 6.1
Proposed HAZMAT information flow.

ID numbers to containers was the best method for tracking HAZMAT through the entire receiving and delivery process. Shipping and Receiving would input descriptive information, barcode ID, initial destination, and responsible party into LTS for each HAZMAT container. This information would then be processed by LTS and LTS would automatically update the inventory. Every time the container is moved or ownership is transferred an audit log would be created.

On May 3, 2005, I attended a meeting with EHS and Shipping and Receiving to discuss the details of importing and exporting needed files from LTS. A staff member requested a modification to LTS that would provide a graphic overlay for the electrical system infrastructure so that emergency personnel would be aware of the exact location of electrical lines in the event of a crisis. During this meeting I discovered that the radioactive waste had been removed from the fourth floor of Founders to a more secure location. The EHS director said that he wanted this system to be adopted by other universities to standardize HAZMAT tracking procedures.

On June 30, 2005, Shipping and Receiving successfully uploaded test information into LTS for HAZMAT materials received. This information contained the proper information on the materials as well as a barcode ID number for each item. Inventory control can now begin at the source of most of the material that arrives on campus and can be tracked as the materials are transported to different areas on campus. This also indicates that uploading data on radioactive materials from the EHS assistant can occur if this action is desired. This allows for all HAZMAT information in LTS to be accessed by anyone who has permission for the HAZMAT functional areas of the system.

7

Security Survey on Campus

SECURITY SURVEY

This chapter illustrates the current state of the hazardous materials (HAZMAT) situation in relation to security and safety at the University. To understand how certain situations evolved, a survey is necessary to record data on current practices for the management of HAZMAT. Observational data gives direction in collecting other data that needs to be obtained and provides information for recommendations to be formulated. The security survey focused on buildings that either used HAZMAT in research or where access to those facilities could be gained through other buildings connected by skywalks or tunnels.

METHODOLOGY OF SECURITY SURVEY

Data was gathered on four Sundays in 2004: May 2, May 9, May 16, and May 23. Sunday was selected because access to buildings is more restrictive than on weekdays and personnel/student traffic would be at the lowest levels since no classes are held on Sundays. My hypothesis was that research laboratories not being used would be secured as would biological elements commonly utilized in research or student activities. The crime prevention coordinator accompanied me on the last two security surveys. For the security surveys I entered C1 and walked along the skywalks through J1, D2, D1, D3, E1, B1, and B2. In addition, I investigated G1 and F1.

DESIGN ASPECTS OF THE CAMPUS

The University was designed to maximize access. The connecting skywalks are a prime example of this architectural decision. With the skywalks, people can navigate unimpeded from one end of campus to another through the interiors of the major structures of the campus (Picture 7.1). Most of the doors designed to seal off corridors were retrofitted to existing facilities. According to a respondent, the doors in C1 were installed after the building had been completed. In the 1960s, D1 was the only structure that existed. As new buildings were constructed, older buildings such as D1 were connected with skywalks so that faculty, staff, and students could easily navigate from one facility to another. Security concerns did not appear to influence the design of the facilities even as recently as 1992, when B1 was designed to have a skywalk from that facility to E1. Improvisations were made by faculty and staff

PICTURE 7.1
Skywalk between D1 and D3.

to offset the loss of research space to student use areas (e.g., classrooms). As seen later in this chapter, a dearth of resources and space led to faculty storing radioactive waste from their laboratories inside a cage in D1 (since removed in 2006) that appeared to be originally designed as a non-HAZMAT locked storage system.

GENERAL OBSERVATIONS

I conducted security surveys independently for the first two surveys. The crime prevention coordinator and I performed security surveys on the third and fourth security surveys. On the third security survey I observed the crime prevention coordinator check access to research laboratories, inquire of personnel as to whether they should be in the facility, and question faculty and students as to what types of specimens were kept in refrigerators. The crime prevention coordinator was carrying a radio and had no visible identification that he was part of the campus security force. On three surveys I was carrying a clipboard and camera, and was wearing a T-shirt, ball cap, and blue jean shorts.

On the fourth security survey, the crime prevention coordinator was wearing a black shirt and a pair of black pants, and was carrying a radio. Again he had no identification indicating he worked for campus security. I was wearing a pair of green army pants, a T-shirt, and a ball cap. In addition, I was carrying a clipboard and two cameras. During all four security surveys the researcher was not asked by anyone who he was or what he was doing in the facilities. On the third and fourth security surveys, the crime prevention coordinator was never asked who he was or why was he performing certain actions (e.g., checking doors for being locked, asking questions about biological elements, etc.). With all four security surveys at no time were any police officers or police guards encountered by the researcher. The only police officer seen was driving a patrol car around campus. On the third security survey, I was inside the facilities for over three hours unobserved and unchallenged. A police officer leaving his shift saw me driving into the parking lot. Although the officer looked directly at me, he took no other action.

Once inside the network of buildings, a person has access to every research building via skywalks and would have had time to complete a number of actions (e.g., steal materials, detonate materials, steal research equipment, etc.). A person also has the capacity to open loading dock gates from the inside to steal large quantities of biological materials or research laboratory equipment. The rooms that contained biological materials did not have a material safety data sheet (MSDS) of present elements posted on the doors of research laboratories. Without this list, first responders and safety personnel would be unaware of the chemical dangers inside, a clear violation of the emergency operations plan (EOP).

SKYWALKS

The skywalks are walkways that connect several buildings together where a person could literally go from one end of the campus to the opposite end without going outside the facilities. The following buildings are connected via skywalks: A1, J1, B1, B2, C1, D1, D2, D3, E1, and K1. Doors to secure research facilities do not exist between buildings, or if they exist, cannot be locked. There are no cameras or surveillance equipment of any type along these skywalks. Picture 7.1 is an external picture of a skywalk bridging D1 building to D2.

A1

A1 is an academic building that houses the School of Social Science and the School of Behavioral and Brain Sciences. A1 houses a School of Behavioral and Brain Sciences class laboratory and several research laboratories. One Geospatial Information Sciences (GIS) research center is also housed in A1. Very little HAZMAT is used in research in this building. A1 is connected to J1 via a skywalk on the third floor and does have an extensive tunnel system for utilities. On Sunday, May 9, 2004, a door was left open on the west side of the building. No camera was in place at the entrance and there was no card reader present. Although this situation is ideal for faculty members to gain access to their offices, it also creates an easement for anyone wanting to gain access through the skywalks to other buildings that had been secured externally. (Also see Table 7.1.)

TABLE 7.1

Summary of A1 Security Survey

Action	May 2	May 9	May 16	May 23
HAZMAT chemicals present	X	X	X	X
HAZMAT radioactive elements present				
HAZMAT biological specimens present	X	X	X	X
HAZMAT chemicals unsecured				
Students present				
External doors open or ajar	X	X	X	X
External door locks not working				
Laboratory doors propped open				
HAZMAT materials in open areas				
Security force member present				
Code key lock door open				
Faculty present				
Researcher/crime coordinator challenged				
HAZMAT materials labeled	X	X	X	X
Cameras present				
Cameras monitored				
Cameras operational				
Inoperable fire extinguisher				
Unlocked laboratory doors				
Access through other building skywalk	X	X	X	X
Unsecured tunnel or utility corridors	N/A	N/A	N/A	N/A
Key code locks	X	X	X	X
Loading dock secured	N/A	N/A	N/A	N/A

Notes: X = yes; N/A = not applicable.

C1

C1 contains a computer laboratory, classrooms, and biology laboratories as well as research laboratories for the School of Behavioral and Brain Sciences. In 2012 C1 was transforming into primarily an administration building for the University but still had research laboratories in one wing of the facility as well as some administrative computing facilities. C1 has a skywalk connection with J1. There is a card reader at the south entrance

PICTURE 7.2
C1 main external doors.

on the second floor. The card reader would not accept my identification card as valid even though I work in C1 (2004). Another person who was neither staff, faculty, nor student was in the building before I gained access through the first floor door on the south side of the facility (Picture 7.2).

The door on the first floor did have a camera viewing the access of the building, but did not have a card reader and did not require a key for entry. The door on the second floor with the card reader did not have a camera in the vicinity of the entrance. On May 9, neither door was unlocked, but C1 was accessible through the skywalk system once I entered A1. The existence of unsecured skywalks defeated the purpose of locking the exteriors doors in C1.

During the third security survey conducted on May 16, I found two alternate ways to enter C1. In addition to the main entrance to the building that has a camera mounted to view accessibility (camera not currently monitored), there is an access point to the basement on the west side of the building (Picture 7.3).

The door was open, which allowed me to enter C1 undetected. The basement area contained the power plant infrastructure to the building as well as laboratory equipment that was labeled "Undergraduate Biology" (Picture 7.4).

PICTURE 7.3
C1 basement door, propped open.

The laboratory equipment was labeled as a compressor and heat sink. Also present were several gallons of chemicals, among them plastic gallon containers labeled “Corrosive 8,” “Sodium Hydroxide,” and “Alkaline.” The basement door was locked on the fourth security survey.

During the week and at night, students are frequently found studying in the main corridors of C1 with the doors to the administrative wing open and unlocked. The research lab doors are locked and the information technology areas are secured by card access.

PICTURE 7.4
Laboratory equipment and chemicals in C1.

PICTURE 7.5
C1 2nd floor external entrance.

On May 2, 2004, a budget office employee was attempting to gain access to the administrative wing when she was stopped and questioned by the crime prevention coordinator. She kept asking the crime prevention coordinator if she was allowed into the building at that time. After verification with the dispatch office (the employee had no university identification with her), the employee was allowed to continue to her office.

The back doors to the building were locked as was the exit near the president's suite. The internal hall doors were locked to the administrative wing on the second floor but were open on the third floor. There is a stairwell within the administrative wing that can enable a person to gain access to the second floor from the third floor, thus defeating the purpose of locking the second floor doors (Picture 7.5). The doors on the third floor were open during the fourth security survey as well.

In addition to security breaches in the maintenance door to the basement and the main door entrance to C1, the door on the second floor with a card reader allowed people to gain entrance up to thirty seconds after another person had used their card (Picture 7.6).

The security breach still existed during the fourth security survey. The doors would automatically unlock for a brief period, allowing a person to gain access to the facility. While I was observing, a student attempted to gain access to the building and was stopped by the crime prevention coordinator to verify her student identification. The key lock to override the door security system had a key broken inside the lock mechanism,

PICTURE 7.6
C1 administration corridor door open.

rendering the override system useless. No camera was located at this entrance. Three students left the facility through this door during my observations. From C1, I can easily access the other research buildings through the skywalks because skywalk doors between the buildings either do not exist or are not closed and locked (Picture 7.7).

These doors appear to be installed for fire containment rather than security purposes. The research laboratories in C1 all had combination key locks on the doors and were secured. However, on the fourth security survey, one of the laboratory doors was propped open with a trash canister. (Also see Table 7.2.)

UTILITY TUNNELS

The utility tunnels connect to all of the University's buildings. Some tunnels are very small, whereas others are large enough to house research facilities. On the third security survey, the access tunnels were all open,

PICTURE 7.7
Skywalk doors.

unlocked, and unguarded. One camera that had been installed by telecommunications was present, but it was not known if anyone was monitoring the entrance to the tunnel. The telecommunication central office switch was accessible through a card reader only. One tunnel exit was locked only by a U-shaped piece of metal. This door could be kicked open. The entrance/exit was located in a concrete open pit covered with a metal grate that could be pried open.

D1, D2, AND D3

D1, D2, and D3 currently house the School of Natural Science and Mathematics (NS&M) and the Office of Undergraduate Studies. These buildings are the oldest buildings on campus, and D2 as well as D3 have been completely renovated in the last couple of years. As discussed earlier, D1 is currently in the process of renovation. The departments within NS&M include geosciences, physics, biology, chemistry, math and science education, and mathematical sciences. There are several research centers based in D1, D2, and D3 as well as the University's telecommunications department and the main PBX telecommunication switch. D1, particularly the second floor, contains several biology laboratories. Refrigerators

TABLE 7.2

Summary of C1 Security Survey

Action	May 2	May 9	May 16	May 23
HAZMAT chemicals present	X	X	X	X
HAZMAT radioactive elements present				
HAZMAT biological specimens present	X	X	X	X
HAZMAT chemicals unsecured			X	X
Students present		X	X	X
External doors open or ajar	X	X	X	X
External door locks not working	X	X	X	X
Laboratory doors propped open				X
HAZMAT materials in open areas				
Security force member present				
Code key lock door open				X
Faculty present			X	X
Researcher/crime coordinator challenged				
HAZMAT materials labeled	X	X	X	X
Cameras present	X	X	X	X
Cameras monitored				
Cameras operational				
Inoperable fire extinguisher				
Unlocked laboratory doors				X
Access through other building skywalk	X	X	X	X
Unsecured tunnel or utility corridors	N/A	N/A	N/A	N/A
Key code locks	X	X	X	X
Loading dock secured	N/A	N/A	N/A	N/A

Notes: X = yes; N/A = not applicable.

filled with biological specimens and chemicals line the hallway. Several of these refrigerators are marked "Not Explosion Proof" (Picture 7.8).

No external thermometers are present on most of the refrigerators and the power supply to these units is unprotected. Other laboratories on the second floor bear signs with HAZMAT indicators, x-ray usage, or lasers in use (Picture 7.9).

On the third and fourth security surveys, access to D1, D2, and D3 was obtained through the skywalk system. In D1, hallways were checked for access to biological elements or research laboratories. During the third

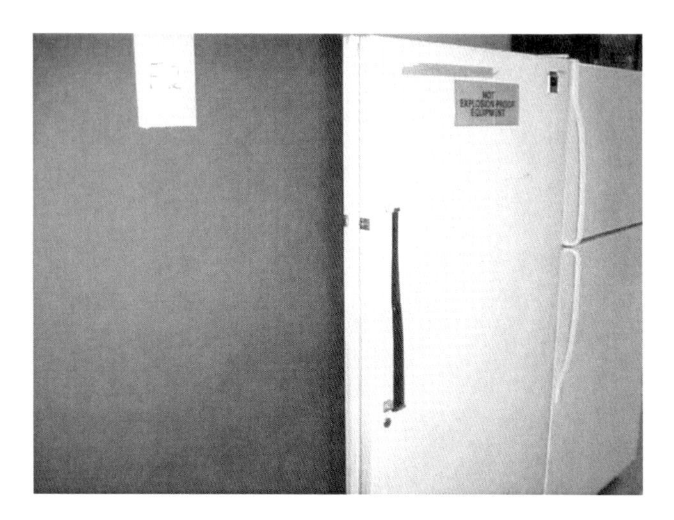

PICTURE 7.8
Biological refrigerators in the main corridor of D1.

survey, one student (possibly a teaching assistant or research assistant) and two employees were encountered. Neither the crime prevention coordinator nor I was asked to identify ourselves or what we were doing in the research area. It appeared that students and faculty assumed we were in the facility for legitimate reasons. During the third security survey, the crime prevention coordinator decided to open what appeared to be a brand

PICTURE 7.9
Warning labels on the refrigerators.

PICTURE 7.10
Refrigerators with external padlocks.

new refrigerator to inspect its contents. The refrigerator was unlocked and packed full of ice. At one point the crime prevention coordinator asked a student, "Who owns this refrigerator?" The student replied that she did not know. At that moment a university employee appeared. At this time the crime prevention coordinator asked the employee, "What are the contents of the refrigerator?" The employee replied that cells and DNA specimens were in the refrigerator. The employee never asked who we were or what we were doing in the area but readily volunteered this information to us. The crime prevention coordinator also asked the employee about locking the refrigerator. The employee responded that this was not possible: "They had not been issued keys yet."

As the crime prevention coordinator and I surveyed the rest of the biology specimens' corridor, we observed that various refrigerators had padlocks retrofitted to the doors and casing of the older units (Picture 7.10).

Neither the crime prevention coordinator nor I was aware of what biospecimens were contained within each unit. No biospecimen labels were present on any of the refrigerators, although various units bore labels marked "Not Explosion Proof." One refrigerator had no visible lock mechanism and was labeled with a radioactive material sticker. Some of the locks appeared to be removable with a large screwdriver or pry bar. During the third and fourth security surveys, it was apparent that several

PICTURE 7.11
Refrigerators without external padlocks.

refrigerators had provisions for padlocks but were not padlocked despite bearing labels such as "Biohazard," "Radiation," or "Not Explosion Proof" (Picture 7.11).

Procedures and practices in accessing such materials are not posted in the corridors where the biospecimens are stored. While surveying the corridor, the crime prevention coordinator locked one of the padlocks on the refrigerator units. A few seconds passed before an employee emerged from his office appearing very annoyed that the refrigerator had been locked. However, the employee did not speak to me or to the crime prevention coordinator. No cameras or surveillance device of any type monitored the area.

The crime prevention coordinator and I then proceeded to the research laboratories on the third floor of D1 that were on the opposite side from where the biology specimens are contained. There is no known centralized inventory of what these laboratories contain. The utility corridor doors on both ends were open, allowing access to interior doors of the laboratories and to three metal canisters filled with pressurized carbon dioxide.

The Iodine Laboratory bore a "Radiation Area" warning label on the door as well as a list of state policies and procedures on the door. However, this door was unlocked and remained unlocked on the fourth security

PICTURE 7.12
D1 external door open with hole in door.

survey as well. The HAZMAT diamond indicator listed the room as a 3 for health (blue), 2 for fire (red), and a 3 for reactive (yellow). The HAZMAT levels on the diamond placard indicate the location of either chemicals or radioactive elements that could be potentially dangerous in the event of an emergency situation.

A second external door on the north side of D1 was found open (Picture 7.12). This door had a hole in it that might have once housed a lock mechanism. The utility corridors and tunnel corridors were open and accessible during the fourth security survey. On the fourth floor of D1 is a radioactive waste depot with several drums of radioactive waste from experiments performed by the biology department (Picture 7.13). The radioactive waste is contained in a mesh wire cage that can easily be unbolted or cut with wire cutters (Picture 7.14). This cage also had an open space between the top of the cage and the ceiling (Picture 7.15). The cage is of similar construction to a cage that exists in the basement of D1 that is used for book and document storage (Picture 7.16 and Picture 7.17).

It is unknown when these cages were constructed but it appears that they were originally designed to store items that were not considered hazardous. The fourth floor of D1 has no security cameras or locks in place to impede access. Furthermore, the top floor has elevator access (Picture 7.18).

PICTURE 7.13
Radioactive waste storage in D1.

PICTURE 7.14
Cage construction.

The external handicap access door on the southwest side of D1 could not be locked unless the power to the door opener was turned off. An external door to the first floor of D1 was found not only to be unlocked, but someone had reversed the lock mechanism so that the door could not be secured (Picture 7.19).

PICTURE 7.15
Space between the cage and ceiling.

PICTURE 7.16
Storage cage used for books and documents.

A fire extinguisher next to research laboratories on the first floor of D1 was expended and unusable (Picture 7.20).

The doors to the skywalks currently cannot be secured because they allow access in each direction at the entrance of D2 and D1. (Also see Table 7.3.)

PICTURE 7.17
Storage cage used for books and documents.

PICTURE 7.18
Unsecured roof at D1.

PICTURE 7.19
Door lock mechanism reversed in D1.

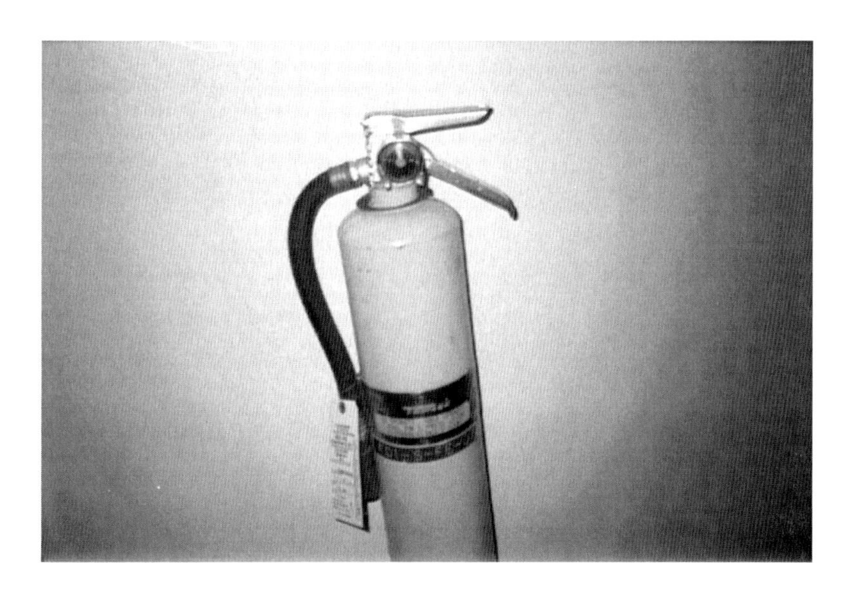

PICTURE 7.20
Inoperable fire extinguisher.

E1

E1 is an NS&M building that contains biology, chemistry, and laboratories for a variety of research programs. Several research laboratories in E1 have recently undergone remodeling. Two cameras are located in E1 but are not currently operable. On the third security survey, the external doors were all secured. On the fourth security survey, the southwest external door was unlocked.

During the third and fourth security surveys, the skywalk enabled both the crime prevention coordinator and I to gain access to the building since no doors are installed on point of entry. One individual was encountered during the third security survey but did not question the crime prevention coordinator or me. All of the research laboratory facilities were secured and locked on the first, second, and third security survey. During the fourth security survey, the main entrance of the nanotechnology laboratory was found open and unlocked. There were no persons present. One faculty member was seen locking and unlocking both his office and laboratory while he moved between locations. (Also see Table 7.4.)

F1

At the time of the original research F1 housed the Environmental Health and Safety (EHS) Department that is responsible for HAZMAT being delivered to and from campus. The building also houses a linear accelerator and several physics laboratories. In 2005, the EHS Department moved to a new set of offices in N1 but still left HAZMAT stored at F1 as of 2009. The doors were tested on the third security survey, and the garage door on the side of the building was found unlocked and accessible. The crime prevention coordinator and I did not enter through the door since the HAZMAT situation for the facility was not known. A radioactive indicator sign was on the fence outside the facility. On the fourth security survey, the main door to the facility was unlocked as well as the door that led to a storage point with a 5 Health/5 Fire/5 Reactive indicator posted. The garage door on the side of the building was again unlocked. An alarm system is in place, but it is not monitored by the University Police Department. (Also see Table 7.5.)

TABLE 7.3

Summary of D1, D2, and D3 Security Survey

Action	May 2	May 9	May 16	May 23
HAZMAT chemicals present	X	X	X	X
HAZMAT radioactive elements present	X	X	X	X
HAZMAT biological specimens present	X	X	X	X
HAZMAT chemicals unsecured	X	X	X	X
Students present	X	X	X	X
External doors open or ajar	X	X	X	X
External door locks not working	X	X	X	X
Laboratory doors propped open	X	X	X	X
HAZMAT materials in open areas	X	X	X	X
Security force member present				
Code key lock door open			X	
Faculty present	X	X	X	X
Researcher/crime coordinator challenged				
HAZMAT materials labeled	X	X	X	X
Cameras present				
Cameras monitored				
Cameras operational				
Inoperable fire extinguisher	X	X	X	X
Unlocked laboratory doors	X	X	X	X
Access through other building skywalk	X	X	X	X
Unsecured tunnel or utility corridors	X	X	X	X
Key code locks	Only on some labs	Only on some labs	Only on some labs	Only on some labs
Loading dock secured	N/A	N/A	N/A	N/A

Notes: X = yes; N/A = not applicable.

G1

G1 was decommissioned around 2008 and demolished in 2009. This research facility was a stand-alone building with no physical connection with other facilities. This building had card readers on outside access points. The facility had two cameras but neither camera was operable. The police department could not monitor the camera feeds from this research facility. The doors were locked and secured. (Also see Table 7.6.)

TABLE 7.4

Summary of E1 Security Survey

Action	May 2	May 9	May 16	May 23
HAZMAT chemicals present	X	X	X	X
HAZMAT radioactive elements present				
HAZMAT biological specimens present	X	X	X	X
HAZMAT chemicals unsecured	X	X	X	X
Students present			X	
External doors open or ajar	X	X		X
External door locks not working				
Laboratory doors propped open				X
HAZMAT materials in open areas	X	X	X	X
Security force member present				
Code key lock door open				X
Faculty present				X
Researcher/crime coordinator challenged				
HAZMAT materials labeled	X	X	X	X
Cameras present	X	X	X	X
Cameras monitored				
Cameras operational				
Inoperable fire extinguisher				
Unlocked laboratory doors				X
Access through other building skywalk	X	X	X	X
Unsecured tunnel or utility corridors	N/A	N/A	N/A	N/A
Key code locks	X	X	X	X
Loading dock secured	X	X	X	X

Notes: X = yes; N/A = not applicable.

B1 AND B2

The B1 and B2 buildings contain classrooms and research laboratories for the institution's engineering programs. During the third security survey, I observed that all research laboratories were locked and were accessible only through card reader locks that were affixed to the doors. On the fourth security survey, the crime prevention coordinator and I were

TABLE 7.5

Summary of F1 Security Survey

Action	May 2	May 9	May 16	May 23
HAZMAT chemicals present	X	X	X	X
HAZMAT radioactive elements present	X	X	X	X
HAZMAT biological specimens present	X	X	X	X
HAZMAT chemicals unsecured			X	X
Students present				
External doors open or ajar			X	X
External door locks not working				
Laboratory doors propped open				
HAZMAT materials in open areas				
Security force member present				
Code key lock door open				
Faculty present				
Researcher/crime coordinator challenged				
HAZMAT materials labeled	X	X	X	X
Cameras present				
Cameras monitored				
Cameras operational				
Inoperable fire extinguisher				
Unlocked laboratory doors			X	X
Access through other building skywalk				
Unsecured tunnel or utility corridors	N/A	N/A	N/A	N/A
Key code locks				
Loading dock secured	X	X		

Notes: X = yes; N/A = not applicable.

not challenged by faculty or students present, and were able to access the loading dock area. The loading dock area had several flammable pressurized gas cylinders present, including oxygen and hydrogen gas cylinders (Picture 7.21). Two cameras were mounted in the loading dock area but were not operational (Picture 7.22).

The clean room was accessible through an emergency exit that had been left open. Behind the building, a liquid nitrogen tank was stored behind a fence surrounded by concrete pillars, which were installed as partial protection from accidental collision.

TABLE 7.6

Summary of G1 Security Survey

Action	May 2	May 9	May 16	May 23
HAZMAT chemicals present	X	X	X	X
HAZMAT radioactive elements present				
HAZMAT biological specimens present	X	X	X	X
HAZMAT chemicals unsecured				
Students present				
External doors open or ajar				
External door locks not working				
Laboratory doors propped open				
HAZMAT materials in open areas				
Security force member present				
Code key lock door open				
Faculty present				
Researcher/crime coordinator challenged				
HAZMAT materials labeled	X	X	X	X
Cameras present	X	X	X	X
Cameras monitored				
Cameras operational				
Inoperable fire extinguisher				
Unlocked laboratory doors				
Access through other building skywalk				
Unsecured tunnel or utility corridors	N/A	N/A	N/A	N/A
Key code locks	X	X	X	X
Loading dock secured	N/A	N/A	N/A	N/A

Notes: X = yes; N/A = not applicable.

On October 13, 2004, I was scheduled to interview Researcher No. 7 at B1 but was prevented from making my appointment because a large amount of chloroform had spilled at the entrance of the laboratory and the hallway corridor. I relocated my interview with Researcher No. 7 to the loading dock outside the B2 building while the Environmental Safety Department cleaned up the spill. (Also see Table 7.7.)

PICTURE 7.21
Empty canisters at B1 and B2.

PICTURE 7.22
Unplugged security camera at B1 and B2.

X1

X1 is a stand-alone building that contains an art gallery and various art studios. According to Staff Member No. 10, X1 was originally constructed to house the physical plant's tractors and heavy equipment. Soon after

TABLE 7.7

Summary of B1 and B2 Security Survey

Action	May 2	May 9	May 16	May 23
HAZMAT chemicals present	X	X	X	X
HAZMAT chemicals unsecured	X	X	X	X
Students present	X	X	X	X
External doors open or ajar	X	X	X	X
Internal laboratory door locks not working	X	X	X	X
Welder present				
HAZMAT materials in open areas	X	X	X	X
Code key locks present	X	X	X	X
Code key lock door open				
Faculty present				
Researcher/crime coordinator challenged				
HAZMAT materials labeled	X	X	X	X
Cameras present	X	X	X	X
Cameras monitored				
Cameras operational				
Security force member present				

Notes: X = yes; N/A = not applicable.

the building was constructed, it was determined that it did not really suit the needs of physical plant and thus the fine arts program converted the building into an art facility for instruction and exhibits. This building stores numerous chemicals that are used in photography, metalworking, ceramics, wood and stone sculpture, painting, and printmaking. On the third security survey, I observed several HAZMAT containers stacked by the back door for disposal. These HAZMAT containers were still in the exact location on the fourth security survey. Since the backdoor is a main entrance into the facility, stacking HAZMAT near the door presents a safety issue. There were twelve one-gallon containers of photo chemicals that had been tagged for disposal and left out in the open (Picture 7.23).

In several of the art studios, chemicals such as lacquers were stored in an unlocked, non-fireproof cabinet. Chemicals such as mineral spirits and paints were kept in various unsecured storage units (Picture 7.24).

Two welders, a pressurized gas type and an arc welder, were in one of the workrooms (Picture 7.25). The welders were not secured and could be used

PICTURE 7.23
Stacked chemicals for disposal at X1.

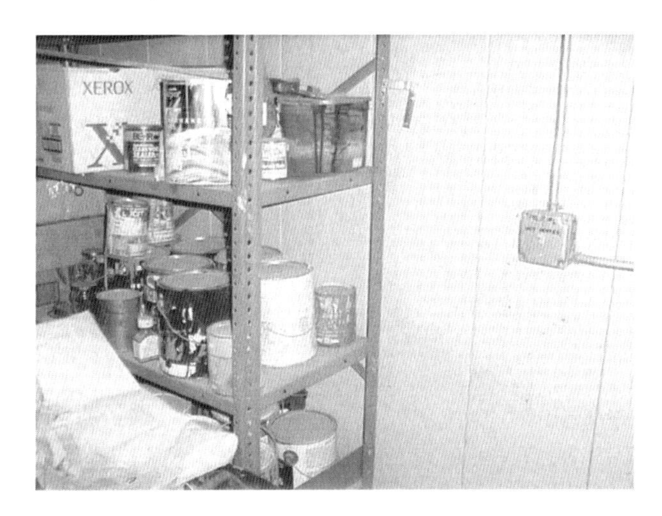

PICTURE 7.24
Chemicals on open shelf at X1.

to break into other facilities. On the third and fourth security surveys, the building was unsecured because two doors by the side porch were propped open and the front door lock mechanism was not functional. On the fourth security survey a student was working in a studio at 5:30 p.m. The facility was supposed to be inaccessible but a door was propped open. The crime prevention coordinator asked the student to please lock the door

PICTURE 7.25
Welder unsecured at X1.

when she completed her work. The student, who appeared very rattled by our presence, responded that she had entered the building through the back door, which was unlocked. She eventually agreed to close the door when she left the building.

During the fourth security survey, the X1 computer laboratory was found unlocked, and there were HAZMAT chemicals present inside (Picture 7.26).

The door had a code-key lock mechanism, but it was not a self-closing door. No cameras were in place at the entrances. (Also see Table 7.8.)

SUMMARY OF SECURITY MEASURES CURRENTLY IN PLACE

Some buildings are equipped with card readers for identification entry. Most buildings still have the traditional key-and-lock system. In C1, the research laboratories have coded entry locks on the doors that restricted access to only personnel who know the code. A1, D1, D2, D3, E1, F1, and I1 have the traditional key-and-lock system on research facility doors, tunnel entrances, and utility corridors.

PICTURE 7.26
Security door open at X1.

TABLE 7.8

Summary of X1 Security Survey

Action	May 2	May 9	May 16	May 23
HAZMAT chemicals present	X	X	X	X
HAZMAT chemicals unsecured	X	X	X	X
Students present				X
External doors open or ajar	X	X	X	X
External door locks not working	X	X	X	X
Welder present	X	X	X	X
HAZMAT materials in open areas	X	X	X	X
Code key locks present	X	X	X	X
Code key lock door open			X	
Faculty present				
Researcher/crime coordinator challenged				
HAZMAT materials labeled	X	X	X	X
Cameras present				
Cameras monitored				
Cameras operational				
Security force member present				

Notes: X = yes; N/A = not applicable.

SUMMARY OF SECURITY MEASURES IN PLACE BUT NOT USED

Several cameras were positioned at entrances. Most cameras were either not working or not being monitored by security forces. Research laboratories in D1 were not locked or, in one case, were propped open with a trash canister (in 2005). There were no security forces verifying that laboratories and exterior doors were locked. C1 had the exterior basement door that was open, unlocked, and unmonitored. The card reader on one C1 door was not functioning correctly and could allow a person to gain access after the door had closed (in 2005). In D1, several of the refrigerators were unlocked or had no lock provision. The utility corridors were unlocked and tunnel doors were open and unlocked.

SUMMARY OF SECURITY MEASURES NOT IN PLACE

Fire alarm exits on doors were nonexistent (in 2005). There was no secure area for biological materials and chemicals that were kept in the refrigerators in D1. Cameras that were in working condition or monitored by the security forces were not present in any of the research facilities that were in working condition or monitored by the security forces. HAZMAT suits and a centralized inventory of chemical agents, biological elements, and radiation detection badges are not available to security forces in any of the research facilities. Lockable doors for the skywalks in D1 and D2 are not in place. Student and employee identifications are not checked, and there is no personnel badge system. In 2012 the refrigerators were located in corridors easily accessible to the general public.

8

Existing Policies and Procedures

GATHERING DATA

After the security survey established security practices in relation to hazardous materials (HAZMAT), I sought existing documentation containing policies and procedures for HAZMAT. The purpose of this chapter is to summarize what documentation was found for chemical agents, biological elements, or specimens used in research. Several departments and faculty were contacted in an effort to obtain the desired information. Many of the policies that were located have not been disbursed to faculty or staff for general use until recently (2005).

METHODOLOGY FOR OBTAINING EXISTING POLICIES AND PROCEDURES

The existing policies and procedures were obtained from the University Police Department. A recent reorganization has placed HAZMAT security under the police department. Previously, HAZMAT security was under the purview of the EHS Department. The University is currently undergoing an evaluation of its emergency operations plan (EOP) as a result of ongoing Federal Emergency Management Agency (FEMA) and Homeland Security mandates.

EVOLUTION OF HAZMAT POLICIES: 2002–2005

In 2002 new operational biological HAZMAT policies were written and approved by the Biosafety Committee. However, those policies were not disbursed to the academic departments or administrative units and thus were not in force until much later (post 2005). In the past, policies and procedures for chemical, radioactive element, and biological specimen usage were the responsibility of each academic department. As of 2005 none of the administrative policies in use addresses inventory control or security of biological HAZMAT materials. Researchers may keep an inventory, but such inventories are not readily accessible nor regularly reported to the administration. The 2004 EOP alludes to an inventory of biological HAZMAT being maintained by the director of Environmental Health and Safety:

> Director of Environmental Health And Safety—Hazmat officer will be notified at the direction of the Chief of Police. This individual is responsible for assisting outside agencies in determining chemicals or substances that may be involved in the emergency. This will be the individual responsible for maintaining Material Safety Data Sheets (MSDS). The MSDS must be made available if requested and are posted outside of any chemical containing area in addition to compliance standards. The fire fighters should not have to enter an area to find out what is in the area.

Currently the EOP addresses biological HAZMAT issues after an emergency has occurred but does not address preventative measures. The only measure that addresses accident prevention is listed under the responsibilities for the director of Environmental Health and Safety.

At present there appear to be no plans or policies being exercised with regard to research activities by the academic departments or research centers for HAZMAT inventory or usage control. Inventory control of HAZMAT is only gathered if a report is due to an external agency. The chief of police asked Police Sergeant No. 1 to investigate safety concerns at Founders Building. On May 27, 2004, Police Sergeant No. 1 attempted to determine if polices existed within the academic departments for HAZMAT controls or procedures in relation to research. As of June 14, 2004, Police Sergeant No. 1 was unsuccessful in finding such policies and did not receive any documents from the researchers. The only reliable way to inventory HAZMAT is when such items are delivered onto campus or when HAZMAT materials are removed from campus. As of 2005, what happened to these elements after delivery and before removal was unknown to the University Police Department or the Office of Environmental Health and Safety. Also there were no protective standard policies in place for biological HAZMAT at the University.

Although the Faculty Senate has approved biological HAZMAT policies and procedures, no faculty member who worked on the policies and procedures was sure when or if the University would implement the policies and procedures (2005). I interviewed the former speaker of the Faculty Senate and two former heads of the University Safety Council. I asked the faculty members a series of questions that can be found in Appendix E. Faculty Member No. 1 had been at the University since 1975 and was charged with ensuring the University had new biological HAZMAT policies and procedures written during his time as the speaker of the Faculty Senate. Faculty Member No. 1 stated that the University Safety Council was formed when the National Science Foundation (NSF) required a committee for chemical, viral, and recombinant DNA research. Originally named the Safety Committee, the University Safety Council coordinated all of the different research committees that oversaw safety. The University "System's" General Council stated that the Safety Committee could not govern all the research committees and that the Safety Committee had to operate independently. Therefore a joint council format was established where all of the research committees had a decision-making authority on the council.

In addition to National Science Foundation requirements, two other events caused the formation of the University Safety Council. The air-conditioning system in G1 malfunctioned and there was no backup plan to deal with this situation. Also, there was no safety manual in existence, a crucial component of any coherent safety plan. The University Safety Council decided to use a safety manual developed by the University of Texas Medical Branch (UTMB) in Galveston that would be reviewed and updated annually by the council. The reason behind the University Safety Council's decision to use UTMB's safety manual is unknown.

Faculty Member No. 1 stated that before the new policies and procedures were written, several smaller booklets served as guidelines on safety. One example was the radiation safety manual. Faculty Member No. 1 stated that often guidelines would contradict what should occur during research activities and that a comprehensive safety manual was needed to provide a uniform guideline. When asked about the oversight of biological

HAZMAT policies and procedures, Faculty Member No. 1 stated that the oversight was in the hands of the deans and individual researchers for each department. When asked about inventory control of biological HAZMAT, Faculty Member No. 1 indicated that this was a problem not only for compliance reasons but also for determining the shelf life of different elements. He also mentioned that when people are not observant and do not communicate with each other, minor situations can become very hazardous. "We were a young university and this [biological HAZMAT] was not a problem for the first 15 years."

Faculty Member No. 1 stated that the latest push for a new safety manual came from the faculty because of the need for uniformity. Faculty Member No. 1 appeared unconcerned about security with regard to biological HAZMAT: "Terrorist acts are just not going to happen here. Drug-making activity is there, but it is not a problem. Having no designated tornado shelters is a much bigger concern." When asked about finding a solution to the biological HAZMAT situation in Founders, Faculty Member No. 1 replied that the new building should solve this situation. Faculty Member No. 1 also stated there was no effective bridge between the University administration and the faculty with regard to safety issues.

I interviewed Faculty Member No. 2 because of his involvement with writing the new safety manual. The safety manual, according to this person, was written to provide flexibility for researchers and allow individual faculty members to be responsible for maintaining their own operational safety. Faculty Member No. 2 also stated that the clean room required safety training on a regular basis for himself and his students. He mentioned that unless the inventory control process was reasonable and almost automatic, faculty would not bother to enter information into an inventory tracking system.

Faculty Member No. 3 had written the safety manual in 1994 when he was the safety chair for the Faculty Senate. Faculty Member No. 3 stated that the policies were written to be more comprehensive and up to date than the policies and procedures currently in place. Faculty Member No. 3 also stated that a powerful dean at that time did not want the Occupational Safety and Health Administration (OSHA) mentioned in the policies and procedures since the state was not required to abide by those guidelines. According to Faculty Member No. 3, OSHA guidelines would have closed the Natural Science and Mathematics machine shop located in the basement of D1 because the space requirements for the type of machinery that was being operated were greater than what existed. The safety manual was written to promote safety while ensuring that the University was not legally bound to adhere to OSHA guidelines. Faculty Member No. 3 added that the University's administration should be properly funding environmental safety programs to ensure biological HAZMAT concerns and safety concerns are addressed.

On June 29, 2004, I contacted two federal agencies to request a copy of their HAZMAT policies and procedures. I was informed that HAZMAT documents could only be released to law enforcement officers, so I forwarded my contact information to Police Sergeant No. 1. Both federal agencies have research materials that are considered restricted or classified and may not release their security procedures to persons outside those agencies due to concerns that releasing such documents could compromise security. On October 1, 2004, the procurement department issued an e-mail to all staff and faculty warning that HAZMAT was not being permitted in the Surplus Warehouse. As quoted in the e-mail from the manager of the Office of Environmental Health and Safety:

> University's Surplus Dept is prohibited from handling all hazardous materials. All laboratory equipment should be thoroughly decontaminated and/or cleaned before Surplus is asked to take possession. They do not have the required means to process hazardous materials. If you have questions in this matter please call the Office of Environmental Health and Safety for assistance. Refrigerators should be cleaned with soap and water and/or decontaminated with 1:10 solution of bleach if they have been used for biological specimen storage. Vacuum pumps must be drained of the oil before the pumps can be sent to Surplus. If radioactive materials have been stored in the refrigerator, after the refrigerator has been cleaned, the appropriate wipe test must be performed to guarantee no contamination is present. Hazardous chemicals must be disposed of through the Office of Environmental Health and Safety. Hazardous material such as batteries (all kinds), corrosive liquids, or paint must also be processed through EH&S. If unsure what procedure you need to follow, please contact the University's Environmental Health and Safety Office for disposal procedures. We thank you for your help and cooperation in these issues.

Two policies that were distributed were the Institutional Animal Care and Use Committee (IACUC) policies and procedures for animal care for research activities and the Institutional Review Board (IRB) policies for research on human subjects. Both of these policies are posted on the

University's Web site for researchers to obtain. The IACUC policies are comprehensive for disposal and use of animals in research. Two areas of concern are not addressed in the policies (as of 2005). The transference of animals from G1 to laboratories is not addressed and allows the researcher to transport animals in any type of container. The delivery of animals to a centralized point on the campus is not addressed. Animals can be delivered directly to a researcher or even a secretary.

A separate policy is posted online for researchers to gain access to G1. This policy is comprehensive not only in gaining access to the facility, but also lists requirements for orientation and training for G1. The Office of the Vice President for Research and Graduate Education is responsible for compliance issues for G1 (as of 2005).

In October 2004, the Staff Council heard a report from the Safety Committee. The committee stated that the Biosafety Manual was still awaiting approval from the deans. The deans' approval has since been given, and the policies have been officially enacted. The EHS director posted the HAZMAT and the biosafety policies online for distribution on June 13, 2005. The general HAZMAT and laboratory policies and procedures are very outdated (1994) and in need of revision.

On July 12, 2005, a faculty member notified me that policies for the Biosafety Committee Charge were posted online at the University. One respondent stated that the committee was not considered active since it had not met in a year. Another respondent further elaborated that a member on the Biosafety Committee was using a chemical that could induce a biological ailment if a human was contaminated with the drug. This person did not notify EHS that the chemical was on campus or that they were using the chemical. EHS discovered its use because the Purchasing Department asked EHS about the purpose of the chemical when they saw it on a requisition sheet. The respondent did not know if the Biosafety Committee had even reviewed a proposal or granted approval to use such an agent that could potentially harm staff, faculty, or students. A second respondent claimed at the time that the researcher did get approval from the IACUC Committee but that the Biosafety Committee was not active at that time and therefore approval was never given by such a committee. When I asked the second respondent about the Biosafety Committee's annual reports, the respondent stated those reports did not exist to their knowledge.

EXISTING HAZMAT SECURITY POLICIES

As stated earlier, security policies for HAZMAT are not mentioned in the EOP (as of 2005). It is commonly understood that each academic department is responsible for preventing the theft or misuse of HAZMAT used in research or instructional settings. During a survey in 2005 of HAZMAT campus procedures, the University Police Department could not find any documented policies in use for handling HAZMAT by the academic departments. The new HAZMAT policies created by the University Safety Council have not been distributed yet to any of the academic departments (as of 2005).

9

Practices and Procedures at Other Institutions

RESEARCH CENTERS AND HIGHER EDUCATION INSTITUTIONS' GLOBAL ISSUES WITH HAZMAT

In previous chapters, the rather unique history of the research site was chronicled to document and provide context for the state of affairs in 2005. Yet the outcome, which includes out-of-date facilities, slow and sometimes glacial reactions to regulatory changes, and make-do solutions in light of tight budgets, is idiomorphic of research-oriented universities largely because of external regulatory and other environmental pressures faced by universities. The purpose of this chapter is to examine these forces and provide examples from a wide range of research universities.

Although each institution has a unique history, current research activities that are funded by federal agencies dictate that each institution must follow set regulations and guidelines for conducting hazardous material (HAZMAT) research. The federal agencies do not take the past history of research activities or current operational procedures into account for institutions that are being audited. The agencies expect that the federal guidelines will be followed for research projects.

HAZMAT at the University consists of radioactive elements, biological specimens (cellular, DNA), dangerous chemicals (in liquid, solid, and gaseous states), and dangerous by-products (waste) produced from research efforts. In many instances, governmental institutions, contracts, or grants mandate the use of hazardous materials in university research. The need to use HAZMAT is complicated by the fact that the research activities within the University are supported by an array of personnel who differ widely in their understanding of the complexity and necessity of HAZMAT guidelines.

FEDERAL ISSUES AND IMPACTS WITH RESEARCH CENTERS

Federal and state HAZMAT laws and guidelines bind both public and private institutions. Guidelines and regulations have been instituted by a variety of federal agencies. For instance, the Drug Enforcement Agency (DEA) regulates controlled substances for experiments on animals and human subjects. The National Institutes of Health (NIH) has guidelines that regulate biological specimens and biohazards for research activities. NIH and the Centers for Disease Control and Prevention (CDC) also have guidelines on how research laboratories must be constructed given the hazard level. The Nuclear Regulatory Commission (NRC) is concerned with radioactive research and the control of radioactive material, some of which can and is used in biological related research. The Environmental Protection Agency (EPA) is focused on biological HAZMAT disposal and usage as it pertains to the protection of the health of students, faculty, staff, and the environment. The Department of Homeland Security (DHS) is now focused on "chemicals of interest," and requires research centers to report those items that are currently used by their organization. The United States Department of Agriculture (USDA) and the Department of Health and Human Services (HHS) are concerned with compliance with the Patriot Act of 2001 (amended in 2002), specifically portions that concern the regulations of biological agents, toxins, and pathogens. HHS also regulates laboratories for health care under the Clinical Laboratory Improvement Amendments (CLIA) of 1988. Research centers that have hospitals or laboratories that treat patients will have waste concerns beyond standard research centers that also contend with chemicals, radioactive isotopes, or select agents. As stated by Thompson (1991, 1):

> And the amount of hazardous wastes that universities need to dispose of is growing. A university hospital alone must dispose of roughly ten pounds per patient per day of human blood, needles, disposable gowns and gloves, body parts, cultures and chemicals.

Other federal agencies have stringent regulations based on contracts or grants that the University is actively researching, for example, an NIH contract.

Federal regulations have been gradually tightened since the early 1980s. Research facilities constructed before the passage of such guidelines either had to be modified in accordance with the new federal guidelines (e.g., Valley Life Science Building at University of California–Berkeley) or taken out of service (e.g., Brookhaven National Laboratory Building 830 Gamma Irradiation Facility) (Department of Energy 2001; University of California at Berkeley 2005). The EPA released information in 2000 relating to several universities that had either been fined for violations or been found to be in noncompliance with EPA guidelines.

> The EPA Regions have found specific examples of noncompliance at universities and colleges that include improperly handling and disposing of hazardous waste materials, boilers and furnaces that do not meet clean air regulations, inadequate monitoring of underground storage tanks, sewage treatment facilities that are not operating properly, and proper abatement of lead-based paint and asbestos. ... Colleges and universities, as well as military educational institutions such as the Air Force Academy and West Point, are required to comply with all applicable environmental requirements like their counterparts in the regulated community to create a safe haven for human health and the environment. Violating the environmental requirements can be costly, for example the University of Hawaii recently paid $1.8 million in civil penalties for violating federal law by poorly managing laboratory waste. (Howell and Hanif 2000)

The EPA has been increasing the number of on-site audits at universities in the last few years. The EPA has fined a number of colleges and universities across the country due to violations of federal guidelines (Associated Press 2001). In 1997, Boston University and Brown University were both fined hundreds of thousands of dollars for EPA violations. As stated by Rene Henry, a governmental relations director for the EPA: "Our inspectors have not been on one campus where they have not found serious problems" (Associated Press 2001).

I contacted the EPA and found that each region of the EPA had its own criteria for auditing facilities to investigate compliance issues (Rebecca Kane, EPA, personal communication 2005).

According to a report by the Department of Health and Human Services, Office of the Inspector General (2004), eleven universities that stored or were going to store "selected agents" were audited to determine if their security procedures were compliant with the 2001 Patriot Act guidelines. The Department of Health and Human Services, in the revised edition, clearly states that facilities that operate with selected agents must adhere to the CDC/NIH biosafety guidelines for laboratories (i.e., *Biosafety in Microbiological and Medical Laboratories* [*BMBL*]). The universities in violation of the Patriot Act responded that CDC/NIH guidelines were only recommended and not required. The Office of the Inspector General then informed those universities that CDC/NIH guidelines for laboratories were in fact binding on their institutions (Department of Health and Human Services 2004). Laboratories and clinical laboratories are under legal statute from the Clinical Laboratory Improvement Amendments (CLIA) of 1988 (U.S. Congress 1988), which dictates minimum procedures and guidelines for laboratory regulation. Laboratories that use materials from the human body for treatment purposes are bound by law to use the guidelines in this legislative statute (U.S. Congress 1988).

Recent work with viruses brought forth new discussion and controversy on developing certain strains of viruses that could be used by terrorists to potentially cause pandemics upon large populations (Fischman 2012). As Fischman states:

> "These lines of work are sufficiently dangerous that they call for independent oversight," says John D. Steinbruner, a professor of public policy at the University of Maryland at College Park; he wants an international-licensing system for researchers who work with dangerous pathogens. (2012, A6)

The apparent lack of government control over biological research has raised new questions on how biological research should be controlled and how biological research centers should be secured as well as kept secret (Fischman 2012). By raising such concerns, different types of research could potentially be restricted to a Los Alamos environment, which is highly restrictive for researchers to operate both within the environment and external to the organization.

In order to create a best practice comparison for HAZMAT procedures and policies, I utilized the same institutions that the University uses for benchmarking purposes. The University has institutions on its list of benchmarked universities that are similar to the University in several characteristics (e.g., enrollment, research expenditures). There are also institutions benchmarked that the University aspires to emulate in the future (e.g., increase in research expenditures.) Since the University was similar in characteristics or was attempting to be similar in areas to the benchmarked universities, the practices, policies, and procedures were investigated to see how those institutions viewed HAZMAT issues. I surveyed nine other public institutions with regard to how their environmental health and safety (EHS) departments are positioned and staffed within their organization. These nine universities had similar characteristics to the University, including the following: number of enrolled students, no law school incorporated into the institution, not a land or sea grant institution, doctoral granting, public institution, a higher amount of research dollars, and campus location (i.e., suburban or urban environment) in relation to the case study university. All nine universities were ranked in the National Science Foundation (NSF) 2002 federally financed research and development expenditures. The University was compared to the following institutions: State University of New York–Albany (SUNY–Albany), Georgia Tech, State University of New York–Binghamton (SUNY–Binghamton), University of Maryland–Baltimore County, University of California–Santa Cruz (UC–Santa Cruz), University of California–Riverside (UC–Riverside), University of California–Santa Barbara (UC–Santa Barbara), University of North Carolina–Greensboro, and University of Wisconsin–Madison. A tenth university, the University of Texas at Austin, was also included even though it is a much larger institution with higher student enrollment and more research dollars than the University. The University typically benchmarks itself against these ten institutions for the purpose of improvement in the areas of academics and research (University of Texas System 2004).

For gathering information on HAZMAT policies and procedures, universities' policies of these comparators were analyzed for content (in 2005). Two institutions that were used as benchmark institutions verified online HAZMAT policies were current and distributed. The other eight institutions did not respond to requests for information and therefore online policies were used for analysis. Some universities did not have comprehensive biological HAZMAT policies and procedures in comparison to their policies and procedures for radioactive elements or chemical agents. Many institutional policies on radioactive material had statements of inventory control for radioactive elements that have been under federal regulations for some time; these policies were used for comparison to biological HAZMAT or chemical HAZMAT policies that were available from other institutions. With the passage of the Public Health Security and Bioterrorism Preparedness and Response Act of 2002, biological HAZMAT policies could potentially begin to proliferate and increase in detail throughout institutions in the same manner that radioactive element policies and procedures gained importance during the Cold War. Additionally, with DHS now mandating reporting chemical inventories to the federal government, there should also be a rise in the number of revised chemical HAZMAT policies as well.

BEST PRACTICES AT BENCHMARK UNIVERSITIES

In 2002, the NSF reported federally financed research and development expenditures for 617 universities (Appendix K). Each of these institutions was bound by federal biohazard guidelines related to research operations. Institutions ranked 151 through 251 (101 universities) averaged $14,371,000 in federally financed research and development. The case study university was ranked in this list with approximately $11,000,000 of federally financed expenditures for research and development. Using federal research and development funds as a proxy for regulatory forces, the institutions utilized in the benchmark study faced, more or less, a similar set of pressures for compliance. It is important to note that three of the institutions within this range were fined for federal EPA violations (University of Hawaii, Brown University, and Boston University). Boston University had two workers hospitalized for exposure in a laboratory with a biosafety level 2 rating incident where scientists were working with a vaccine (Berger 2005).

In 2004 the Sunshine Project, a private international organization, surveyed 400 institutions' institutional biosafety committees (IBCs) to gather data on university compliance with NIH guidelines (Field 2005). Many of those surveyed are the same aforementioned institutions. The Sunshine Project is attempting to have more policies and procedures emplaced to control biological weapons (Field 2005). The survey was gathering information on the effectiveness of institutions' biosafety committees that oversee research performed with biological agents. As stated by Field (2005):

> Ultimately, 82 percent of the public universities and 79 percent of private universities replied to the request—a far better response rate than among biotechnology companies, 45 percent of which replied. However, only two-thirds of public universities and 57 percent of private universities polled produced minutes containing protocol review, and a number of universities refused to answer the select-agents question. While the federal government

> does not bar institutions from disclosing that information, universities are not legally obligated to do so.

The Sunshine Project (a nonprofit watchdog organization) has been sharply critical of lax procedures of university IBCs.

> "They like to think of themselves as being the nice-guy enforcer," Mr. Hammond said. "Now is the time to use the stick instead of the carrot. It was time to use the stick years ago." The NIH's Mr. Shipp acknowledges that the relationship between the NIH and the institutional biosafety committee is largely based on trust. The agency does not collect IBC minutes to confirm that they are reviewing research, and it does not require biosafety committees to certify that they are in compliance, as it does with institutional review boards. (Field 2005)

Physical containment and laboratory safety have been the IBC's primarily focus in an institution's biosafety policies and procedures (Field 2005). Security and inventory control of biological agents appear not to have an equal weight of urgency that safety and physical containment have currently with IBCs. Aside from a potential audit that might be performed by the Department of Health and Human Services, Office of the Inspector General, federal standards for all institutions for the approval of biological research and ensuring compliance of biological research do not exist.

GENERAL CHARACTERISTICS OF INSTITUTIONS' ENVIRONMENTAL HEALTH AND SAFETY OFFICES

All of the institutions benchmarked had their respective EHS offices report to the administration line of reports rather than to the academic line of reports (Table 9.1). They varied on whether the department reported directly to the vice president/vice chancellor level. Some departments were under the physical plant or facilities department responsibilities. The number of personnel typically varied between ten and thirty-seven employees. The highest level of authority in these departments is at the director level.

TABLE 9.1

General Characteristics of Other University EHS Departments (2005)

	Georgia Tech	SUNY–Albany	SUNY–Binghamton	UC–Riverside	UC–Santa Barbara	UC–Santa Cruz	UM–Baltimore County	UNC–Greensboro	UT–Austin	UW–Madison
NSF ranking of federally financed R&D expenditures at universities and colleges (2002)	42	224	123	137	87	135	145	301	24	9
Top Position, Director		x		x	x	x			x	x
Top Position, Safety and Health Manager							x			
Top Position, Associate Director			x							
Top Position, Associate Vice President for Facilities	x									
Top Position, Associate Vice President for Administration										
Reports to Vice President of Finance and Business		x								
Reports to Vice President of Administration			x							
Reports to Senior Vice President of Administration and Finance	x									
Reports to Facilities and Planning Management										x
Reports to Vice President for Campus and Employee Services									x	
Reports to Vice Chancellors for Administrative Services or the Office of Business and Administrative Services				x	x	x				
Reports to Department of Facilities							x	x		
Number of Staff	13	9	11	19	30	11	N/A	7	35	37

Notes: x = applicable to the institution; N/A = not applicable.

TABLE 9.2

EHS Personnel Ratios for Tenured Faculty and Federally Financed Research Expenditures

University	Number of EHS Personnel	Number of Tenured Faculty	2002 NSF Federally Financed Research Expenditures	Ratio of EHS Personnel to Tenured Faculty	Ratio of Expenditures to EHS Personnel
Georgia Tech	13	545	$165,680,000	1:42	1 per $12,744,615
SUNY–Albany	9	377	$40,497,000	1:42	1 per $4,499,667
SUNY–Binghamton	11	302	$8,959,000	1:27	1 per $814,545
University of California–Riverside	19	419	$32,305,000	1:22	1 per $1,700,263
University of California–Santa Barbara	30	658	$78,370,000	1:22	1 per $2,612,333
University of California–Santa Cruz	11	370	$32,901,000	1:34	1 per $2,991,000
University of Maryland–Baltimore County	N/A	261	$29,376,000	N/A	N/A
University of North Carolina–Greensboro	7	342	$3,340,000	1:49	1 per $477,143
University of Texas–Austin	35	1433	$219,158,000	1:40	1 per $6,261,658
University of Wisconsin–Madison	37	1555	$345,003,000	1:42	1 per $9,324,405

Table 9.2 lists the institutions by the number of tenured faculty (Integrated Postsecondary Education Data System 2004), the 2002 NSF-reported federally financed research and development expenditures, the ratio of EHS personnel to tenured faculty, and the ratio of EHS personnel to federally financed research and development expenditures. The information in Table 9.2 appears to indicate that Georgia Tech, the University of Texas at Austin, and the University of Wisconsin–Madison have

understaffed EHS departments according to research expenditures for 2002. The University of North Carolina–Greensboro and SUNY–Albany would also appear to need additional staffing based on the ratio of EHS personnel to tenured faculty.

Characteristics of Environmental Health and Safety Departments at Benchmarked Universities

Each institution used as a benchmark by the University has structured its EHS department within the organization in ways that reflect how each institution places a different level of priority on the importance of the EHS department. Some institutions have their EHS department report to a director in the physical plant or facilities. Other institutions appear to place a greater importance on their EHS department by having it report to an associate vice president or vice president.

At Georgia Tech, EHS reports directly to the Associate Vice President for Facilities, who in turn reports to the Senior Vice President of Administration and Finance (Edward Guida, director of Environmental Health and Safety, Georgia Tech, personal communication 2004). According to Lee Zacarias, environmental health coordinator at Georgia Tech (personal communication 2005), Georgia Tech regularly inspects laboratories at least once a year, and any research that does not have approval is terminated until the proper committee gives authorization. Georgia Tech is about to begin construction of three to four biotech buildings and a stem cell research center with new institutional focus on biomedical technology. Georgia Tech is in the initial phase of tagging chemicals with barcodes for a cradle-to-grave inventory control process. Due to safety concerns, Georgia Tech currently stores its inventory of select agents on removable media and not on an electronic database system. According to Zacarias, many attempts to gather practices and procedures from other schools have not been successful. Furthermore, Zacarias has stated that Georgia Tech personnel have been reviewing potential technology solutions for the last two and a half to three years before deciding upon a software package called Groove.

The University of Maryland–Baltimore County is the second example of an EHS department under the facilities management department. The EHS department has a safety and health manager for the top position (University of Maryland–Baltimore County 2004). (In the absence of an organizational chart, this was inferred because the duties performed by the Facilities Department include contacting the Office of Environmental Health and Safety.)

The third institution studied that appears to operate EHS as an extension of a facilities department or physical plant is the University of North Carolina–Greensboro. The Office of Safety reports to the Department of Facilities, which in turn reports to the Office of Business Affairs. The department has seven employees, one of whom is a professional firefighter. Most of the current staff members hold college degrees that specialize in environmental health and safety (University of North Carolina–Greensboro 2012).

The University of Wisconsin–Madison was the fourth institution studied that placed its safety department under a facilities department (University of Wisconsin–Madison 2012). Considering that the University of Wisconsin–Madison has medical facilities on campus, it is surprising that the institution does not integrate its safety department at a higher organizational level or have the department report directly to the Vice Chancellor for Administration. Currently the department appears to be grouped with other physical plant operations. The institution appears to have a highly trained and educated safety department staff. The safety department works closely with the institutional biosafety committee, department chairperson, and college administrators to ensure researcher compliance (Klein 2005). The biosafety officer at the institution stated that problems were virtually nonexistent, but there were methods to exercise controls if required (Klein 2005). One method that the institution can use is withholding the release of awards or restricting the researcher from ordering animals for experimentation (Klein 2005).

SUNY–Albany has recently placed more emphasis on its EHS department than in the past. The department is no longer an extension of the physical plant and plays a more integral role with research activities at the institution. At SUNY–Albany the director reports directly to the Vice President for Finance and Business (Franconere 2004). This reorganization is apparently fairly recent since this department once reported to the Assistant Vice President of Facilities who in turn reported to the Vice President of Finance and Business (Franconere 2004). According to Director Vincent Franconere (personal communication 2004), the reorganization has allowed his department to function more effectively and efficiently. The EHS department at SUNY–Albany is composed of nine employees, including the director (SUNY–Albany 2004). In the future it will be interesting to see how many institutions begin to move EHS departments away from physical plant or facility department operations, and begin to transition EHS functions to a higher level within the organization.

Several institutions have their EHS departments report to the Vice President for Business Affairs or the Vice President for Administrative Services. The safety departments at the University of California–Santa Barbara, University of California–Santa Cruz, and University of California–Riverside report to either their respective Vice Chancellors for Administrative Services or the Office of Business and Administration Services, and are headed by a director. The safety department at the University of California–Santa Barbara has more than thirty people who have received highly specialized training for environmental health and safety concerns The annual report for the EHS department cited an increase in activity with the advent of citations and fines being levied against various departments at the university (University of California–Santa Barbara 2012).

SUNY–Binghamton appears to view the EHS department and the police department primarily as safety organizations rather than physical plant or business service operations. SUNY–Binghamton has its EHS department report to the Director of Public Safety, who in turn reports to the Vice President of Administration. Most employees hold degrees that specialize in environmental health and safety areas (SUNY–Binghamton 2012). Another unique approach to locating EHS organizationally is the University of Texas at Austin. The University of Texas at Austin's EHS department reports directly to the Vice President for Employee and Campus Services (University of Texas at Austin 2012). Many of these institutions are located near a large body of water (e.g., Pacific Ocean, Lake Michigan) and have specific needs for their EHS departments to perform. Institutions such as the University of California–Santa Barbara have a Campus Diving Safety Officer in their EHS department (University of California–Santa Barbara 2012). Not every institution has a need for that type of program or personnel. The University of Wisconsin–Madison has to focus a great deal of attention on issues involving medical waste, more so than other universities that do not have that type of waste on their campuses. Every institution will need to have EHS correctly staffed for the needs of the campus programs, and they will need to have EHS at a high enough level organizationally to have a positive impact on the institution.

RESEARCH UNIVERSITY HAZMAT GUIDELINES

Based on my investigations on the University's HAZMAT procedures and policies, I investigated what other high quality research universities required for safety at their institutions. When I began my research on HAZMAT policies and procedures in 2004 and 2005, I initially searched for standards on institutional Web sites, only to experience difficulty in locating these practices for several institutions. In addition to the benchmark universities described earlier, I reviewed the policies and procedures of several universities that were ranked above the University on the NSF federally financed research and development expenditures for 2002. A summary of the benchmarked institutions' HAZMAT policies for 2012 is referenced in Table 9.3 and Table 9.4.

In 2004 to 2005, when I initially reviewed the benchmark universities, policies, some of the university policies that were posted online appeared to be out of date, although more current policies could very well exist that are not publicized for a variety of reasons. In 2004 to 2005 the policies and procedures that were posted on institution Web sites appeared to be focused on reacting to HAZMAT incidents rather than developing preventative measures to enhance security or safety or refer to HAZMAT safety and security in general terms. When initially reviewed in 2004 to 2005, most institution policies that were publicized online that I reviewed did not list any procedures on how to maintain an inventory of hazardous substances, how to protect such items, or how first responders are to contend with hazardous items in the event of an emergency. There were some institutions that listed provisions for employee accountability for misconduct, fraud, or misuse of HAZMAT.

When I revisited the HAZMAT policies of the benchmarked universities in 2012, most of the policies and procedures were posted online for the communities' knowledge and use. Policies that were missing or unavailable in 2004–2005 were no longer so in 2012, indicative of a considerable effort on the universities' part to post comprehensive HAZMAT policies online. There were still instances where the benchmarked universities did not go into too much detail about inventory procedures or security for HAZMAT. The policies and procedures appeared to focus more on general safety guidelines for research efforts.

TABLE 9.3

HAZMAT Policies and Procedures for Institutions, Part 1

	Georgia Tech	SUNY–Albany	SUNY–Binghamton	UC–Riverside	UC–Santa Barbara	UC–Santa Cruz	UM–Baltimore County	UNC–Greensboro	UT–Austin	UW–Madison
NSF ranking of federally financed R&D expenditures at universities and colleges (2002)	42	224	123	137	87	135	145	301	24	9
Focus on laboratory safety	x	x	x	x	x	x	x	x	x	x
Reference state and federal laws	x	x	x	x	x	x	x	x	x	x
Reference CDC/NIH biosafety guidelines	x	x	x	x	x	x	x	x	x	x
Reference NIH recombinant DNA guidelines	x	x	x	x	x	x	x	x	x	x
General biohazard guidelines	x	x	x	x	x	x	x	x	x	x
Blood pathogen guidelines	x	x	x	x	x	x	x	x	x	x
Security of biohazard	x	x	x	x	x	x	x	x	x	x
Reference to Patriot Act	x					x				
Reference to Public Health Security and Bioterrorism Preparedness and Response Act of 2002 (*BMBL* 5th edition)	x	x	x	x	x	x	x	x	x	x
Provisions for disposal of HAZMAT or medical waste	x		x	x	x	x	x	x	x	x
Inventory of biohazard (selected agents)	x		x	x	x	x	x	x	x	x
Audit of laboratories	x		x	x	x	x	x			
Biohazard governed by office of compliance	x	x	x					x		
Governed by institutional biosafety committee	x	x	x	x	x	x	x		x	x

Notes: x = policy posted online.

TABLE 9.4

HAZMAT Policies and Procedures for Institutions, Part 2

	Georgia Tech	SUNY–Albany	SUNY–Binghamton	UC–Riverside	UC–Santa Barbara	UC–Santa Cruz	UM–Baltimore County	UNC–Greensboro	UT–Austin	UW–Madison
Chemical HAZMAT policies	x	x	x	x	x	x	x	x	x	x
Radiological HAZMAT policies	x	x	x	x	x	x	x	x	x	x
Waste disposal HAZMAT policies	x	x	x	x	x	x	x	x	x	x
Homeland security mandates for HAZMAT posted online						x		x	x	
Emergency management guidelines	x		x	x	x	x		x	x	x
Laser laboratory safety guidelines and policies	x	x	x	x	x	x	x		x	
Reference to Homeland Security Chemical Facility Anti-Terrorism Standards 2007				x		x		x	x	
Security of chemicals	x	x	x			x			x	
Inventory of chemicals	x	x	x	x	x	x		x	x	x
Security of radiological elements	x	x	x	x	x	x		x	x	x
Inventory of radiological elements	x	x	x	x	x	x	x	x	x	x
References to FEMA policies and procedures				x	x	x				
Chemicals governed by office of compliance or committee	x			x	x				x	x
Radiological elements governed by office of compliance or committee	x	x	x	x	x	x		x	x	x
Controlled substances policies and procedures		x	x	x	x	x			x	x

Notes: x = policy posted online.

For physical security of facilities, best practices are primarily concerned with restricting general access that does not involve HAZMAT issues specifically. As stated by Peter Hamilton (1987, 47):

> The purpose of a barrier is physically to prevent access, or to secure premises tightly against entry by burglars, to keep honest employees from areas they have no need to enter and to restrict movement to predetermined routes. The nature and strength of the barriers will differ according to the requirement, and will range from thick walls with solid doors, to bullet-resistant transparent screens. When planning a security system, therefore, it should be borne in mind that the barriers form the primary protection. A well-planned system utilizes the concept of defence in depth, normally three lines or stages: the first at the site perimeter, the second at the areas within the building, and the third at the perimeter of sensitive areas within the building. Depending on the risk, detection system will operate on any or all of the defence lines, and will be wired to allow for the use of different types of detection device.

Although Hamilton refers to generalized protection schemes for facilities, any detailed information on protection of facilities largely depends on the purpose of the building, organizational tolerance of security measures, and the manner in which the building was constructed. Best practices with regard to physical security would be to apprehend persons at the first stage of protection, the site perimeter.

At Binghamton University, State University of New York, biohazard procedures are referred to the institutional biosafety committee. In 2005, the university did not have general guidelines on disposal of medical waste, audit on laboratories, and general safety guidelines (Valcik 2006). However, by 2012 Binghamton University had comprehensive policies and procedures for all HAZMAT that had been posted online (Binghamton University, State University of New York 2012).

The University at Albany, State University of New York, had HAZMAT policies (referred to as guidelines) supervised by the Office of Research Compliance (Valcik 2006). The university has since posted comprehensive policies and procedures online for its institutional use (University at Albany, State University of New York 2012).

Georgia Tech has comprehensive biohazard policies that address transference of biohazardous materials to other institutions, approval of biological experiments, responsibility of the principal investigators for security of their laboratory, and policies that incorporate procedures to comply with the Patriot Act along with governmental regulations on selected agents (Valcik 2006). Georgia Tech also had provisions in the biohazard policies for access to selected agents and a risk assessment provision for experiments (Valcik 2006). Since 2005 Georgia Tech has updated its HAZMAT policies online and incorporated the latest federal regulations for chemical inventory and chemical security (Georgia Tech 2012).

The University of North Carolina–Greensboro references biohazard NIH guidelines on recombinant DNA and states that the Office of Research Compliance is responsible for ensuring all federal guidelines are followed (Valcik 2006). The University of North Carolina–Greensboro has increased the amount of information as well as policies and procedures for HAZMAT issues on the Internet (University of North Carolina–Greensboro 2012).

The University of Wisconsin–Madison references the American Society for Microbiology practices and procedures, NIH guidelines for recombinant DNA, and the CDC/NIH biosafety laboratory manual (Valcik 2006). The university also has a general provision for security and inventory control for laboratories and dangerous biological elements (Valcik 2006). The University of Wisconsin–Madison has increased the amount of HAZMAT policies and procedures online, but there was no mention of chemical inventory in reference to the Department of Homeland Security Appropriations Act 2007: Section 550, DHS-2006-0073, RIN 1601-AA41, 6 CFR Part 27–Chemical Facility Anti Terrorism Standards (University of Wisconsin–Madison 2012). There was also no policy to be found on policy or procedures for laser usage found online on the environmental health and safety Web site (University of Wisconsin–Madison 2012).

The University of Maryland–Baltimore County has biohazard policies that focus on laboratory safety and does have a good reference for labeling laboratories for hazard levels (Valcik 2006). The university also has a provision in the HAZMAT policies for having each item with a material safety data sheet (MSDS; Valcik 2006). Since 2005, the University of Maryland–Baltimore County has increased the number of policies and procedures significantly and has improved the amount of information available to their organization (University of Maryland–Baltimore 2012).

The University of Texas at Austin references the CDC/NIH biosafety guidelines, NIH recombinant DNA guidelines, Texas blood-borne pathogen rule, and OSHA blood-borne pathogen rule (Valcik 2006). The University of Texas at Austin has kept and updated HAZMAT policies and procedures for its university (University of Texas at Austin 2012).

The University of California–Riverside references generalized biohazard guidelines that mention compliance with federal and state laws (Valcik 2006). The University of California–Santa Cruz references the CDC/NIH biosafety manual, NIH recombinant DNA guidelines, and has a bloodborne pathogen program (Valcik 2006). The University of California–Santa Barbara has biohazard policies and procedures primarily focused on safety of laboratories (Valcik 2006). All three institutions have upgraded their HAZMAT policies and procedures to address the new federal and state local statutes (University of California–Riverside 2012; University of California–Santa Barbara 2012; University of California–Santa Cruz 2012).

BEST PRACTICES AT HIGHLY RANKED RESEARCH INSTITUTIONS

In addition to the benchmarked institutions, I investigated the HAZMAT practices of several research universities that were ranked in the NSF's 2002 list of federally financed research and development expenditures. The benchmarked institutions and other institutions not only maintain information on HAZMAT standards that the University should consider adopting, they also provide additional practices that can improve the University's general safety standards. Most universities that referenced biological HAZMAT online referred to the existing CDC/NIH standards for laboratory construction for both biosafety levels 1 to 4 and animal biosafety levels 1 to 4 (2005). Some universities had faculty-led committees that formulated and enforced regulations and standards (2005). Most of these institutions also required their researchers to request clearance to conduct biological research from the biosafety committee prior to the onset of research activity (2005). These biosafety committees frequently require audits on laboratory facilities (2005).

Several universities maintained strict HAZMAT policies and procedures regarding inventory control of radioactive materials (as of 2005). One university required that all purchases of radioactive isotopes be carefully tracked through the university from purchase to final delivery. Although radiological element control was strictly regulated at several universities, these same institutions did not address other aspects of HAZMAT, particularly for biological elements, with the same level of control. Most universities were concerned with biological and chemical safety practices in the laboratory rather than institution-wide security concerns or inventory control. One university required strict inventory control over controlled substances only. Other universities had detailed policies and procedures for the proper disposal of HAZMAT materials. Since 2005 the benchmarked universities have put more emphasis on the new statutes that relate to security and inventory control for chemicals.

In 2005, I interviewed a former assistant director for the environmental health and safety department at the University of Texas Southwestern Medical Center (UT Southwestern) after I requested biosafety policies and procedures but never received the documentation. The respondent (now the new manager for EHS at the University) was originally authorized to provide information to me from UT Southwestern according to the manager of EHS at that institution. The former assistant director informed me that UT Southwestern maintained inventories on select biological pathogens and adhered to the 1988 CLIA guidelines for laboratories that perform medical tests on patients. The center's laboratories are constructed to CDC/NIH standards and many of the laboratories are constructed to biosafety level 3 standards. Animals are transported within a pneumatic tube system. International students who request access to selected agents undergo background checks to ensure they are legally permitted to have such access. However, the medical center experienced difficulty in checking international students for the exportation of technology as defined by current federal guidelines.

SUMMARY OF APPLICABLE UNIVERSITY HAZMAT POLICIES

With a couple of exceptions (i.e., Georgia Tech), inventory control and security procedures that have been reviewed lack detailed information about when and how biological HAZMAT should be inventoried, how biological HAZMAT should be safeguarded, and who should create and maintain the inventory of biological HAZMAT (2005). Institutions typically leave responsibility for inventory control to the principal investigator. The United States Air Force controls its inventory of radioactive elements by assigning identification numbers to containers (U.S. Air Force 1997), a procedure that universities can adapt for their own use in controlling biohazard containers.

Initially the federal government provided no adequate safety standards in its CDC/NIH biosafety guidelines for the proper handling of biological HAZMAT in university laboratories, shortcomings that the federal government attempted to address in the latest biosafety guidelines. There are three possible reasons that make adequate security standards exist for HAZMAT issues and facilities. The first reason is that each federal agency has set its own detailed safety criteria that would not suit the unique research conducted at university laboratories. The second reason could be that safeguarding HAZMAT is a new concern and that there has not been a centralized incentive to develop operating procedures for HAZMAT issues. Even the Department of Homeland Security provides only vague guidelines for the protection of HAZMAT (as of 2005). The third reason could be that it is difficult to outline procedures to protect research laboratories in the highly accessible campus environment with buildings that have been constructed at various times in the institution's history that in many cases predate current infrastructure guidelines. For example, a research facility built in the 1930s may not have the ability to be wired for electronic locks and surveillance cameras due to a lack of crawl space above ceilings or under floors. Therefore, best practices will vary for each type of research within each type of facility. As stated by Philip P. Purpura (1989, 6):

> Losses from crimes, fires, accidents, natural disasters and so forth, are the obvious problems. The strategies to counter these losses are numerous. A reader with even the slightest knowledge can list the more common ones: security officers, alarm systems, planning and preparation, and the like. It is of the utmost importance, however, for practitioners to look beyond these basic strategies to new methods. A never-ending search for new and better ideas is a necessity in dealing with our complex, changing world.

The CDC/NIH research laboratory specifications address general security for biological research levels 3 and 4 and animal biological research levels 3 and 4 while providing some very basic guidelines for security (CDC/NIH 2004). Best practices will be those practices that most successfully conform to the organization's functionality and facilities.

In the United States, regulations for the construction and maintenance of research universities' biological laboratories rely on guidelines from the CDC, NIH, state laws, and local fire codes. In Canada bioresearch laboratories are regulated by Health Canada, a federal department that specifies laboratory designs for construction and establishes guidelines for laboratory procedures (Health Canada 1996). Several institutions maintained extensive documentation for inventorying, controlling, protecting, and disposing of radioactive materials, but there was less focus on chemical or biohazard materials.

Very few biohazard policies mentioned the Patriot Act or the Public Health Security and Bioterrorism Preparedness and Response Act of 2002, although several policies did reference the CDC/NIH biosafety manual for laboratory policies and procedures. Several of the biohazard policies did reference the CDC/NIH biosafety manual or World Health Organization (WHO) biosafety manual for policies and procedures for laboratories. When compared to the benchmarked universities, the University appears to have an understaffed EHS department and lacked effective biohazard policies (as of 2005). University policies seem to lack adequate controls and inventory mechanisms for biohazard compared to the extensive policies and procedures for radiological elements. Although the Biosafety Committee Charge document does state that the Biosafety Committee is to develop procedures for maintaining a catalog for biohazardous materials and pathogenic agents, it does not specifically define "catalog" as inventory control. This statement is very vague and unclear on exactly what the committee is authorized to develop for policies or procedures. Since 2005, the University has posted a link to the CDC/NIH *BMBL* 4th edition, but currently this link directs the user to the main CDC homepage since the *BMBL* 5th edition is the current version. Furthermore, according to respondents (interviewed in 2005), the Biosafety Committee had not met in over a year and does not review contract or grant proposals that contend with biological elements. One respondent stated that there are two outside members who are supposed to be on the committee but had yet to attend a meeting (2005).

The structure of EHS departments in relation to the organization of universities varied from academic representation of faculty on an institutional biosafety committee to a more centralized role for EHS departments. For example, some institutions have their EHS departments maintain centralized chemical inventories. Other institutions expect the principal investigators to keep inventories of the chemicals or biospecimens that exist in their laboratories. Some EHS departments have personnel specifically educated or trained in industrial hygiene or toxicology. Other institutions staff their EHS departments with physical plant workers or administrators. In 2005, the University hired an EHS manager who had a master's level degree in toxicology and another employee with a graduate degree in industrial hygiene. This represents a shift of focus from the previous incarnation of the office that was staffed by physical plant employees. However, turnover occurred and by 2012 the University replaced an interim director with a director who held a doctorate degree.

An EHS department's function is largely dictated by the responsibilities faculty have for their own research-centered areas. Faculty will be able to interface with administration more effectively if the EHS department is staffed with personnel knowledgeable in scientific methods, research, and elements. Otherwise faculty will have no reason to respect administrative oversight of research operations to ensure compliance with state and federal guidelines.

In addition, security needs have increased with the additional federal requirements to secure select agents. This will lead many institutions to determine who will be responsible for security for these types of elements. With universities that address the issue of security, the policies reviewed indicated that principal investigators were responsible for security of their research areas. What were not addressed in the reviewed policies were standardized security mechanisms for laboratories and buildings. At the University, the university police department is tasked with security for HAZMAT. This creates overlaps and gaps of responsibilities between the EHS department, the university police, and the faculty member who is performing the research. Some security and safety measures are sporadically dispersed at the University and do not uniformly cover all areas of research. For example, security cameras are currently the responsibility of the department, which may or may not allow first responders access to the video feed. Furthermore, there are no standards in place for the type of cameras to be used in facilities. Many institutions have their EHS departments report to the business affairs line of business (Figure 9.1). Without faculty support, tensions will arise between the business affairs line of business and the academic affairs line of reports. Faculty will not appreciate personnel from business affairs dictating how research operations should be performed if those personnel are not qualified in scientific methodologies.

With the exception of the vague statement of developing procedures for maintaining a "catalog" in the Biosafety Committee Charge at the University, a provision for inventory control of biological elements was not covered by any reviewed policies (in 2005). There were no provisions for identifying containers of waste or vials of different biological specimens for tracking purposes. Although principal investigators may be slated to have responsibility for items in their laboratory, they do not have control

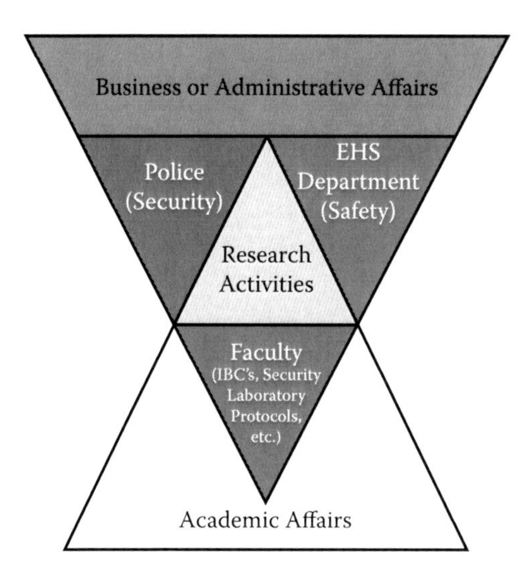

FIGURE 9.1
Interaction between business affairs and academic affairs.

of items either being delivered or exported (e.g., medical waste) from their research area of control. Several policies for other institutions were reviewed that did have practices and procedures for identifying containers of radioactive elements or waste. Though the policies did not refer to a centralized database system to track these items, such a system utilizing a unique identification number for each container would make it possible to track the status of every container that held radioactive elements or waste by-products. Biological elements could be inventoried in a similar method.

As of June 13, 2005, the University posted new general laboratory safety and biosafety policies; however, these policies are still outdated compared to other universities' biosafety policies. Most of the general safety policies date to 1994 and the biosafety policies need to be revised to comply with new federal legislation, such as the Public Health Security and Bioterrorism Preparedness and Response Act of 2002. Currently there are only provisions in the general safety policies for the University Safety Committee to enable research faculty and administrators to revise and update policy. The 1994 general guidelines seem to reflect a perceived need for a radiation safety officer but not for a biological safety officer. Several other universities policies that were reviewed provided for a biological safety officer. On the basis of the evidence gathered from other universities, especially those which have more experience, I recommend that the University staff the EHS Department as shown in Table 9.5 (2005).

TABLE 9.5

Proposed Staffing Requirements for EHS at The University, 2005

Position	Currently Staffed	Responsibilities	Requirements or Proposed Credentials
Safety director	x	Overall responsibility for EHS duties and compliance	Master's in toxicology
Chemical safety officer	x	Responsible for inventory, safety, auditing laboratories with chemicals, and governmental compliance for chemical elements	Master's in industrial hygiene
Radiation safety officer	x	Responsible for inventory, safety of radioactive materials, auditing laboratories with radioactive material, and governmental compliance for radioactive materials	Extensive training in containment, safety, cleanup procedures, and accidental exposure to radioactive material
Biological safety officer		Responsible for governmental compliance, OSHA standards, auditing all laboratories with biological agents, and audit inventory of potentially dangerous or select biological specimens	Ph.D. in a field of biology or M.D. preferred
Fire safety specialist		Responsible for compliance of all fire and building codes, first responder status for emergencies	Certified firefighter
Environmental manager	x	Responsible for waste disposal and compliance to all EPA regulations	Extensive training in HAZMAT disposal and EPA regulations
Inventory control specialist		Responsible for all database management, data entry for inventory control, and mapping of infrastructure	Bachelor's in computer science and Geospatial Information System (GIS) certified
Risk manager		Responsible for safety training, CPR training, driver training, and development of risk management procedures	Knowledge of OSHA standards and risk management practices
Technology safety officer		Responsible for the safety standards, inventory, and governmental compliance with lasers, x-ray machines, and other high-technology equipment	Bachelor's degree in engineering required; knowledge of EPA and NRC regulations

(continued)

TABLE 9.5 (CONTINUED)

Proposed Staffing Requirements for EHS at The University, 2005

Position	Currently Staffed	Responsibilities	Requirements or Proposed Credentials
Laboratory compliance officer		Train personnel and students in general laboratory safety; ensure laboratory facilities are built to CDC/NIH standards	Bachelor's degree in engineering
Administrative assistant	x	Administrative support for EHS	High school diploma

As shown in Table 9.5, the University's EHS Department had five full-time employees. Four staff members handled many of the proposed positions' roles. There are too many government regulations for only four personnel to effectively oversee compliance for an entire research university. If compliance is the EHS Department's responsibility, then the office needs to be staffed in the appropriate areas. However, compliance is only one responsibility. Training and crisis response is also a mission of the EHS Department. The appropriate resources are needed for effective response to occur. Currently the University has one person serving as the biosafety officer, the radiation safety officer, and the laser safety officer. The complexity of the current rules and regulations governing biological agents dictates that one staff person who is highly knowledgeable with containment and safety of biological specimens must be dedicated for that particular specialty. A person who holds a doctorate in biology or an M.D. would not only have the knowledge to perform the duties as assigned but also could interact more effectively with researchers who experiment with biological elements.

The fire safety officer should ensure that firefighting equipment is current and in compliance, and inspect facilities after construction or renovation to ensure that fire and building codes have been followed. A fire safety officer who is certified as a firefighter also provides the campus with an onsite first responder for emergency situations.

A laboratory safety officer would be responsible for ensuring that the University was in compliance with all OSHA and CDC/NIH guidelines on the construction of new laboratories or laboratories under renovation. The laboratory safety officer would also be responsible for training employees and students in proper laboratory safety protocol as well as developing policies and procedures for general laboratory safety. The laboratory safety officer would be responsible for ensuring that all policies and procedures are updated and revisited on a regular basis.

A risk manager would be added to the current EHS staff to establish a risk management policy and procedure manual. The risk manager would ensure that workplace OSHA standards, with the exception of bloodborne pathogens, are being complied with for multiple departments (e.g., physical plant workers.)

A technology safety officer would be responsible for the safe operation of high-technology laboratory equipment such as lasers and x-ray machines. As rapidly as technology is progressing, other equipment that is used in research may need safety oversight in the future. The broad job title of technology safety officer provides flexibility in assignment of future laboratory equipment.

An inventory control specialist would be added to the staff to provide support in data entry and programming efforts. Centralized information systems that can track radioactive materials, chemicals, and biological specimens require a database administrator to support both the server and the application. The inventory control specialist would be responsible for ensuring that the inventory information is input into the system correctly and that the information is accessible to personnel. If the inventory control specialist were also trained in Geospatial Information Systems (GIS) applications, then he or she could effectively map infrastructure such as sprinkler systems, waterlines, and electrical systems for each building, and provide additional information for first responders and emergency personnel to contend with emergency situations.

Other institutions appear to rely heavily on the principal investigator to draft HAZMAT security procedures. This reliance does not encourage any uniform security standards for the institution, which in turn can result in serious security gaps. Without centralized auditing on security procedures, it is difficult, if not impossible, to ascertain if new federal biosafety guidelines are being followed. Relying on procedures set by the principal investigator does not protect the institutions from liability or protect the principal investigator from civil lawsuits or criminal charges in the event of an incident. The policies should reflect inventory control; auditing procedures; security guidelines; and compliance with the Patriot Act, the Public Health Security and Bioterrorism Preparedness and Response Act of 2002, Department of Homeland Security Appropriations Act 2007 (Section 550, DHS-2006-0073, RIN 1601-AA41, 6 CFR Part 27–Chemical Facility Anti Terrorism Standards), OSHA standards, export control of technology guidelines, and the CLIA Act of 1988. As of 2005, all of these provisions were absent from not only the University's posted biosafety policies but from other institutions' posted policies as well.

From an organizational standpoint, I recommend that the EHS department report to a vice president for administration. If that position does not exist, then I recommend that the EHS department report directly to the president. To interact with the EHS department, a Biosafety Committee should be appointed by the provost to advise EHS on new policies and procedures when new guidelines or techniques are employed since bio-HAZMAT are relatively new compared to radioactive isotopes. Currently the president appoints the Biosafety Committee members. The biosafety officer from EHS should serve ex officio on the committee. This allows the faculty to have input with EHS on how EHS interacts with faculty members affected by new mandates. The Biosafety Committee should also be responsible for approving new research proposals before a grant proposal can be submitted, hold meetings quarterly, and keep minutes for documentation on each meeting. Proposed structuring of Biosafety Committee is illustrated in Figure 9.2.

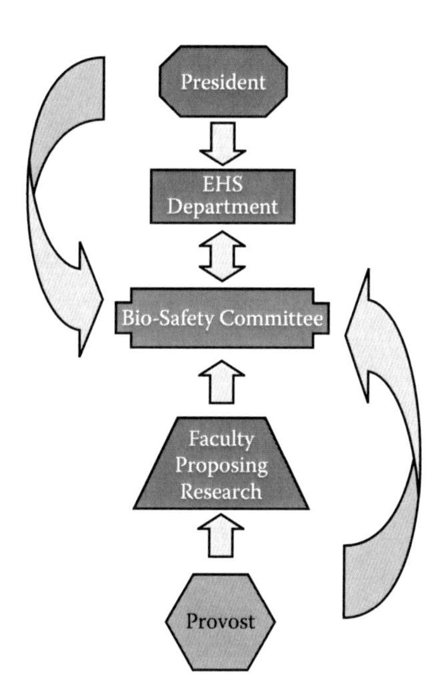

FIGURE 9.2
Proposed organizational operation of biosafety units.

With the increase in federal regulation, EHS will become more important than ever for research universities wishing to either expand federally funded research or to retain the federal funding they currently enjoy. The new structure of the Biosafety Committee would allow for the provost to have input to EHS procedures by appointing faculty members to the Biosafety Committee.

In summation, this chapter has provided a number of different policies, procedures, and practices from not only the University's benchmark institutions but also provides examples of practices from top-tier research universities around the country. Where federal standards are required for HAZMAT policies, a minimum required level of operational procedures and practices must exist. Some institutions, such as the University, are above the minimum federal requirement in some areas in HAZMAT and below those minimum standards in other aspects. The federal policies and procedures have been improved upon by a number of universities (e.g., University of Wisconsin–Madison) to address the increasing concern over HAZMAT. However, some issues, such as inventory control and security, can be improved further for safety and security of HAZMAT in general.

PROCESS OF MUTUAL ACCOUNTABILITY FOR BIOSAFETY

The Biosafety Committee and EHS will need to cooperate and share information. If EHS performs an audit on a biological laboratory, those results should be made available to the Biosafety Committee. If a professor wants to use a drug that can induce a biological condition in a human or animal, EHS should be notified. If both areas are made accountable for safety and security of biological HAZMAT, then the likelihood of policies and procedures being followed by researchers will increase. There is no oversight with regard to research activities with the exception of the IACUC (as of 2005). The result is that individual researchers are free to establish standards but, in some instances, without apparent documentation or approval from EHS or the Biosafety Committee (2005). The point of mutual accountability is to provide all parties involved with a consistent and clear set of guidelines that are in compliance with relevant federal and state standards. The Committee and the EHS have a separation of duties in terms of creating and overseeing parts of the research process but are mutually accountable for the outcomes.

10

Summary, Recommendations, and Concluding Remarks

The purpose of this research was to examine one university's policies and practices with regard to the procurement, use, storage, and disposal of hazardous materials (HAZMAT) in the context of a changing internal organizational structure and a changing regulatory environment. Essentially, this research is a case study of one research institution (described as "the University") that focused on the relationships among the institution's unique history, its environment, and its place in the evolving nature of federal research that led to the existing conditions of its HAZMAT. The University, originally a research think tank, evolved into a public research university with an enrollment of thousands while simultaneously experiencing a substantial increase in the quantity and complexity of its research. The University's environment transformed from an isolated, rural area into an expanding suburban city. Federal regulation of HAZMAT exerted greater influence over the University's activities as the surrounding area filled with homes and private businesses, placing citizens much closer to facilities where hazardous biological materials were stored and used. The degradation of adequate public funding for the University's growing enrollment and research needs was mirrored by the degradation of the University's primary science building as more laboratories were added without the necessary upgrades to comply with new federal regulations. The organizational culture also underwent change as instructional responsibilities were added to the faculty's research duties. Adequate security became difficult to maintain as increased foot traffic due to the construction of new classrooms carved from old laboratory spaces resulted in students walking through corridors once restricted only to researchers.

Although research universities may have unique histories and workflow processes with regard to HAZMAT, it appears that environmental pressures eventually force research institutions to standardize operations to meet regulations. Change will occur at different rates for each individual institution due to each university having its own specific research operations and unique faculty characteristics that contend with HAZMAT. The types of HAZMAT used for research will determine the risk factors for noncompliance with government regulations. To ensure compliance, many institutions will need to change the structure and processes of the organization to provide adequate support for researchers using HAZMAT. Fundamentally, universities need to protect critical physical assets, information, and materials while also ensuring that researchers can have the freedoms they need to engage in creative endeavors. At their normative extremes, the excesses in security could prove to be overly bureaucratic and stifling, while at the other extreme, individual laboratory controls can break down, resulting in misplaced materials and altered experiments. Two factors influence current and future policy. First, the scientific ethos and method provide a powerful directive for the researcher to desire control over his or her laboratory. This plus local policy was the predominate means by which science advanced in biological sciences until the later 1800s.

Increasingly, the federal government, in response to perceived problems, threats, or crisis, has passed legislation that allows federal agencies (e.g., National Institutes of Health [NIH]) to create guidelines that impact universities' ability to perform research operations. The rise of federally sponsored research programs began with the NIH in 1930. Over time, more federal agencies became involved in research programs related to health, safety, and security for biological elements. The Drug Enforcement Agency, which was established in 1973, extended its jurisdiction over controlled substances used in research. The development of new technologies spurred more regulations and more oversight by federal agencies. The Atomic Energy Commission (AEC) is an example of a new agency created specifically to oversee the proper use and handling of radioactive material. The latest federal department that has instituted new HAZMAT guidelines for research centers is the Department of Homeland Security (DHS) with the passage of the Department of Homeland Security Appropriations Act 2007: Section 550, DHS-2006-0073, RIN 1601-AA41, 6 CFR Part 27–Chemical Facility Anti Terrorism Standards.

With federally sponsored research programs came federal regulations, first with federal guidelines from the NIH and later from a multitude of federal agencies, universities began facing an increasingly complex and changing regulatory environment. At the same time, more sensitive and potentially hazardous materials made their way into research projects. Even so, behavior patterns for many universities did not embrace changes in policies, procedures, and practices that were perceived as a possible infringement to research activity. This was particularly true for those institutions with small research efforts and those lacking already established administrative structures. This created two main concerns for universities: new HAZMAT regulations that rendered existing science facilities obsolete and new laboratory procedures that caused personnel trained in older methods and practices to become noncompliant.

Thus, a multitude of factors—the increasingly federalized research agenda and the regulatory strictures attached, the presence of research universities in urban areas, the increasing perceived virulence of research materials, and the rise of the new animal anti-vivisectionists and their fellow travelers as well as sporadic but effective public outcries—pressed for change and dictated the need for striking a balance between HAZMAT security sufficient enough to assuage public concerns and protect assets while simultaneously encouraging independent thought, experimentation, and the sharing of results.

ORGANIZATIONAL THEORY IN THE CONTEXT OF THE RESEARCH

The investigation into the evolution of the University's HAZMAT condition utilizes several organizational theories. By utilizing the research approach known as grounded theory, existing HAZMAT conditions can be analyzed to determine why and how these conditions evolved within the University. Life cycle theory indicates how organizational drift could occur as the amount of research performed at the institution outpaced available monetary resources and why the loci of control shifted away from investigator-controlled research. When the University expanded in size and research capabilities, more federal regulations forced the organization to change existing policies and procedures to remain compliant with federal guidelines. The research addresses how organizational drift led to the misalignment of the University HAZMAT policies, practices, and procedures away from current federal regulations due to the

University's expansion. The University was unable to accommodate many of the recent federal guidelines because existing resources were applied toward developing its research capacity while new resources were funneled into expanding classroom instruction. Eventually this shortcoming would cause a shift from investigator-controlled research to a more organization-centered research model. Finally, agency theory is relevant for explaining another element of how organizational drift and decoupling caused misalignment with current federal regulations. Because the University was originally a research think tank, a logic of confidence in the researchers existed in the culture of the organization that did not provide oversight to the researcher's activities. Administrators and faculty assumed that researchers had enough knowledge about their work to operate safely in research laboratories. As research expansion at the University outstripped existing resources, the researchers were left to work on their projects with little or no resources to support their existing operation, much less accommodate new federal requirements on an existing infrastructure that was not going to be upgraded.

SUMMARY OF THE RESEARCH METHODS

Qualitative methods such as conducting security surveys and personnel interviews, researching best practices, and analyzing archived HAZMAT documentation were used to gather data throughout the duration of this research. To construct a robust HAZMAT timeline that adequately explained the existing HAZMAT situation for the University, it was necessary to use these different sources to triangulate the data. Best practices were collected through documentation of other universities' policies and procedures, guidelines from federal agencies, and interviews of personnel employed at other universities. By analyzing this documentation, it can be determined if the University's current operations meet the minimum standards of other entities. The participant observer research method was utilized to gain insight on operational aspects for biohazards. By working with administrative units that exercised jurisdiction over HAZMAT policies and procedures at the University, additional data was obtained that would have been difficult to obtain independently. Furthermore, the participant observer method enabled the implementation of some of the recommendations proposed in this study that will improve safety and security at the University.

Data collected from these research methods were critical in formulating a grounded theory from which the topic could be analyzed. Initial data gathered through unobtrusive observations enabled the formulation of survey questions that would determine how and why the current HAZMAT conditions existed. It was necessary to determine if the facilities that contained HAZMAT had always operated according to certain methodologies or did the existing situation evolve from one that once was more stable and secure. (The data gathered in the security survey would reveal that the situation had, in fact, evolved not only due to a research-centered institution adding instruction to its mandate but also to the inclusion of research programs housed in facilities that were not originally constructed to contain them.)

SUMMARY OF RESEARCH ISSUES AND FINDINGS

There are several internal and external pressures that affect the organization, the greatest of which involve the federal government. Federal agencies, such as the Department of Health and Human Services, Office of Inspector General, are now aggressively auditing research universities for compliance with federal guidelines and finding that many research universities have compliance shortcomings. Research universities that had traditionally permitted researchers to conduct their research operations with limited oversight have now increased requirements that researchers must follow so that they can continue their endeavors. Compliance with new requirements requires research universities to invest more resources in administrative support such as the establishment or expansion of an environmental health and safety department. Failure to comply with federal mandates not only opens institutions to the threat of fines or other punitive action from federal and local agencies but can also result in catastrophic accidents or criminal activity at research universities.

The University's new role as a teaching institution with an increasingly diversified research presence has compelled the University to undergo organizational change to meet new internal and external operating pressures. Increased federal legislation prompted the University's administration to improve safety and security of HAZMAT by hiring a new director and additional staff for the Office of Environmental Health and Safety, commissioning an inventory control software application capable of tracking HAZMAT elements through the University from location to location or from person to person, and removing from D1 the radioactive waste that had been generated from various biomedical experiments.

The administration has become aware of shortcomings revealed in the security surveys conducted for this research. There was a noticeable lack of security forces present in the surveyed buildings and an unwillingness of staff, faculty, or students to challenge unidentified individuals who enter research areas. Skywalks, which were a great convenience to the campus community, also permitted greater accessibility to research facilities from other buildings and inadvertently created a situation whereby outsiders could remove or destroy the contents of containers located within common areas of these facilities. A number of laboratory doors were left unlocked, and there were no working security cameras in any of the research areas that were monitored by university police. To forestall future incidents, the administration has begun to investigate ways to address these security issues while balancing the needs of researchers to conduct their experiments without undue restrictions on their activities.

Since the conclusion of this investigation, there has been a reinvigorated effort toward updating existing policies and procedures at the University. The University's HAZMAT policies, which for years had been outdated and not widely distributed, have now been updated and distributed throughout the institution's research community. The Institutional Animal Care and Use Committee (IACUC) policies and procedures have been distributed and been consistently revised since the late 1970s. The emergency operations plan (EOP), which had been previously distributed to the public, is currently under revision by the University Police Department.

In spite of their unique origins and development, all research universities eventually adopt similar practices for the proper handling of HAZMAT due to increased federal regulations imposed on all institutions. Therefore, it is useful to examine best practices at selected universities to determine industry and governmental standards that could be useful at the University. The following recommendations are based on this evaluation.

CONCLUDING OBSERVATIONS

The topic of HAZMAT in a research university context is currently receiving a great deal of attention, especially with the rise of terrorist activities and the subsequent passage of recent federal legislation such as the Patriot Act. Although biological HAZMAT has been an issue with research universities for decades, many institutions do not appear to place the same emphasis on biological HAZMAT as they do on radiological or chemical elements. Consequently, institutions have lagged in the development of adequate policies and procedures to regulate biological HAZMAT and have thus become decoupled with regard to this issue. Mistakes, accidents, and criminal activity are the internal and external forces that can prompt a system to take a tighter coupling approach on biological HAZMAT activities. It can be dangerous to permit organizational drift in an institution that is involved in research activities, which can have far-reaching consequences for the campus population and people who live in surrounding areas.

For research centers in general, there are more biological research laboratories being constructed as a biosafety level 4 laboratory than have existed previously. Plum Island Animal Disease Center, for example, is being decommissioned and a new replacement laboratory at biosafety level 4 is being constructed at Manhattan, Kansas (Department of Homeland Security 2009). Biosafety level 4 laboratories have a number of security and safety features that are being incorporated into the new laboratory designs. These features include advanced security, and air filtration and decontamination systems that are state of the art (Pappalardo 2009). Such technology is needed along with sound policies and procedures to prevent accidents from occurring. New technology, such as the HAZMATID Ranger from Smith's Detection, offers research centers new capabilities to potentially detect HAZMAT that may pose a problem for research centers ("Product of the Month" 2009). For research centers, biometric access can potentially be a solution for certain security issues to gain access to facilities or laboratories if the organization has the resources and will to implement such a security system. With any type of system that is installed, it is still critical for the organization to remember that training is crucial to maintain sound business processes in the research environment. No matter how sound a solution may be, if the personnel executing or implementing that solution are not trained, or do not have the will or the skill sets necessary for successful implementation, the solution will fail.

It is important that an institution maintain a willingness to embrace new organizational forms and new techniques through mutual accountability of the researchers and administration to achieve greater safety and security over its biological HAZMAT materials. The measures and techniques that have been proposed in this document could enhance security while maintaining research productivity. The application of existing technology can improve physical security by preventing the theft or destruction of assets and harm to personnel. This in turn prevents the disruption of research activities thereby improving, not hampering, research productivity. An investigation of the organizational structures of other universities indicates that the existence of a strong and well-staffed environmental health and safety department can serve as a valuable partner for faculty and researchers to ensure that research activities are compliant with federal and state laws. However, compared to benchmark institutions, the University's environmental health and safety department was too understaffed to manage a campus of over 14,000 students and an increasing level of research (in 2005).

As of 2005, the University has acknowledged this by hiring a new director and additional personnel, although more personnel might be required in the future. The University is also addressing security needs by hiring additional police officers for the University Police Department, another support unit that was considered understaffed. Safety and security at D1 has been a continuing issue since the construction of the building and the advent of the biology program. In the case of D1, inadequate financial resources and a dearth of adequate space had led to improper storage of biospecimens in common corridors and of radiation hazardous waste on an upper floor. As of 2012 most of D1 had been renovated to house class laboratories and there was only one wing of the building that still contained refrigerators for experiments and storage in the hallways, which is a good sign of progress. With continued remodeling of D1 and the addition of two new research facilities, this should permit the University to decommission D1 as a science building, which will in turn alleviate many of the shortcomings discovered in the original research in 2005.

Although there is still much that needs to be accomplished, the University had already made progress in improving the HAZMAT situation. Perhaps the greatest accomplishment will be the successful shift in perception of the importance of properly maintaining HAZMAT. Mutual accountability of researchers and administrative departments can foster cooperation and create improvements in the safety and security of HAZMAT while permitting research endeavors to flourish at the University.

RECOMMENDATIONS

General Security Measures for Facilities Containing HAZMAT

The University can benefit from safeguarding HAZMAT and research laboratories from internal and external threats. Some research universities (e.g., the University of Wisconsin–Madison) address security for their research facilities in their policies. By utilizing additional security measures to protect the most sensitive research areas, the University can enable security personnel to focus their efforts toward campus crime prevention while preserving open access to nonresearch areas of campus. Preventative measures can also be employed to ensure limited access to sensitive research facilities when classes are not in session.

Although the University has already improved security by hiring additional security forces and by installing security cameras in an additional building, there are other areas that could benefit from additional measures that could make HAZMAT facilities more secure. The creation of a standing committee composed of academic and research faculty, a police representative, and a representative from the environmental health and safety department can evaluate current security practices and policies, and institute new measures that could improve security. This committee could assist in streamlining aspects of security by reviewing policies and practices to access research facilities, upholding security camera standards, and formulating a mechanism whereby individuals involved with research projects can be granted special authorization to those projects. The committee can also keep the University updated on new federal and state regulations to ensure compliance and evaluate private industry methods to verify if these methods can be integrated into the University's existing security measures.

Construction of Research Facilities Housing HAZMAT

The University is progressively upgrading research facilities to Centers for Disease Control and Prevention (CDC)/National Institutes of Health

(NIH) standards for biosafety as well as Occupational Safety and Health Administration (OSHA) standards—widely utilized by many universities (see Chapter 9)—with the addition of new facilities and the decommissioning of D1 as a research facility. To fully comply with CDC/NIH regulations specifically for biological agents, it is ideal to isolate new facilities from other university facilities and to ensure that they are not connected by skywalks or tunnels so that casual student traffic through these facilities can be kept to a minimum. The University's researchers and Environmental Health and Safety Department can evaluate the current state of existing research facilities and upgrade those facilities to proper standards when budgets permit. For more sensitive research facilities, a biometric security system could be installed in place of card readers or physical keys. This would be one added step to ensure that the person who is entering the facility is the authorized person.

The development of a master plan for construction or renovation of biological laboratories could enable the University to construct biosafety level 3 laboratories that in turn could permit researchers to apply for contracts and grants that specify that type of laboratory, thus leading to more research revenue for the University (2005). A master plan for biological research facilities could enable the University to develop a "science complex" that can house animal research facilities and commonly shared research equipment, which could maximize the efficient use of space and equipment and reduce human contact with the research animals. The University of Texas Southwestern Medical Center (UT Southwestern) currently has biological research facilities organized in a "scientific complex" for maximum efficiency of biological research (former assistant director for environmental health and safety, UT Southwestern, personal communication 2005).

Biosafety Committee

Many research universities (see Chapter 9) maintain biosafety committees that approve research proposals, establish procedures, and provide oversight for ongoing research projects. The University has the mechanism in place for a biosafety committee that can provide the same services at the institution. The biosafety committee has the capacity to update the University's biosafety policies to address mandatory federal statutes, provide a mechanism for peer oversight in various fields of scientific research, and explore the possibility of upgrading existing biomedical research laboratories to the Clinical Laboratory Improvement Amendments (CLIA) 1988 standards for hospitals and clinical laboratories (2005). UT Southwestern abides by CLIA 1988 standards (see Chapter 9). If the University were to follow CLIA 1988 standards, its laboratories would automatically follow a national industry standard that is designed to improve the health and safety for all faculty, staff, and students who work in those facilities. The biosafety committee can also investigate a blood-borne pathogen guideline for research laboratories (see Chapter 9) and a policy for the transference of animals to and from research facilities similar to that developed by UT Southwestern (former assistant director for environmental health and safety, UT Southwestern, personal communication 2005).

Inventory Control

In response to new federal regulations that hold research universities accountable for how select agents, biological elements, chemicals, and controlled substances are used, stored, and disposed, it is recommended that the University use an electronic database of HAZMAT that can enable a department to track what type of HAZMAT exists by container in certain areas, and to inform all safety and security forces of potential hazards in the event of an incident. Arizona State University, for example, currently has an inventory software application that tracks chemicals from initial delivery to disposal (Grayson 2005). The same method could be used to track specific biological elements.

To enable this database to function at an optimal level, the University could use barcodes, radio-frequency identification (RFID) tags, or GPS locators to track items that are disbursed by the receiving department to research facilities. The ID number (for whichever system is used) is adhered directly to the HAZMAT container and is assigned attributes that create an audit trail, enabling personnel to track changes in location and ownership. This ID system should be modeled after laboratories in hospitals that use barcodes to link the contents of containers to the correct patients. The University of California is considering tracking cadavers with barcodes or RFID devices to safeguard them against theft or loss (Locke 2005). In light of incidents at other research institutions (e.g., missing plague vials from Texas Tech University), these methods can also be used to safeguard potentially dangerous biotoxins within the University. Other methods that could be used would be to employ RFID tags or GPS locators on certain high-value or high-risk items that need to be accounted for in an electronic database for inventory control purposes.

Epilogue

THE ENVIRONMENTAL HEALTH AND SAFETY DEPARTMENT

Shortly after I gave the chief of police the security and safety surveys in 2005, the University hired a new environmental health and safety director. The new director had previously served as an assistant director at a prestigious research medical university and brought expertise in compliance issues, knowledge of chemical safety, and knowledge of microbiology/biological safety. Positive action items were implemented at the Environmental Health and Safety (EHS) Department that had previously not been seen during the research. As of 2007, the EHS Department (including the director) grew from two employees to eight employees. As of 2012, the department included the following employee functions:

Employee #1—Assistant Vice President, Chief Safety Officer
Employee #2—Worker's Compensation and Risk Management Administrator
Employee #3—Assistant Director, Laboratory Safety Manager
Employee #4—Fire Marshal, Life Safety Coordinator
Employee #5—Deputy Fire Marshal, General Safety
Employee #6—Environmental Program Manager
Employee #7—Emergency Management Coordinator
Employee #8—Office Assistant

INVENTORY CONTROL

In 2005, I started working with the director of EHS at the University to establish an electronic inventory in LTS for chemical, radiological, and biological elements. The initial plan was to have the shipping and receiving department bar code and enter the elements into their shipping and receiving system, and then upload the information into LTS for tracking purposes. Although the new shipping and receiving system was successfully developed, the business processes that would enable the successful transmission of the information into LTS could not be resolved, and the new system was abandoned in 2006. The new system would have allowed for information to be added to LTS in real time without a delay in updating information. The new system would have also integrated the use of hand held scanners, making the new system much more robust than the previous tracking system. As of 2012, the information must be entered into LTS manually because there is no method to import that information via scanning or through the shipping and receiving system. Due to turnover at the top EHS position, the effort has lost traction. Although an older list of chemicals have been input into LTS, newer updates of chemical inventories has not been maintained.

For establishing LTS and the HAZMAT tracking procedure, the University received an Award of Recognition for the Unique and Innovative Safety Program from the National Safety Council–Campus Safety Health and Environmental Management Association in 2006. The University had not received that award prior to 2006.

HAZMAT POLICIES

In 2005, the EHS director posted the HAZMAT policies online and publicized the new revised policies. In 2006, the EHS department at the University constructed an Exposure Control Plan for Bloodborne Pathogens and posted those policies online. As stated in Chapter 6, these policies were in limbo before the new director was employed in 2005. Additionally, policies and procedures were posted online by the EHS department in regard to controlled items, controlled substances, emergency eyewash/shower program, laboratory safety inspections, laboratory design review, laser safety, materials safety data sheets, radiation material and laser usage, radiation safety, and shipment of hazardous materials. These policies and procedures represent a significant positive shift in how the EHS department is operating at the University.

COMMITTEES

Since the new EHS director has been employed, the Institutional Biosafety Committee (IBC) has been meeting regularly, and the policy memo that governs the IBC has been updated twice since 2006. This reflects another positive shift of direction. Before, the memo governing the IBC had not been revised since 2002. Additionally, the members appointed to the IBC are easily found on the University's Web site. Before 2005 this information was difficult to obtain.

SAFETY AND SECURITY SHORTFALLS ADDRESSED

When I started my research in 2004, there were some troubling issues that were occurring at the University in regard to physical safety and security. Soon after I submitted my research findings, the University began taking steps to rectify these issues. The new EHS director ordered removal of the radioactive waste drums that were stored in the mesh wire cage (see Chapter 7, Picture 7.12). In 2005, the University purchased and modified an existing building that would house research (the N1 facility) from D1, which would allow D1 to be reduced in function as a science building. In 2007, the University opened a new research facility (H1) complete with high-technology laboratories and safety equipment to house existing and new research. Some professors transferred their research to H1. This in turn has cleared out many of the refrigerators that lined the corridors of D1 (see Chapter 7, Picture 7.7). The items that were stored in corridors are now being stored in specifically designed storage facilities in both N1 and H1. In addition the new buildings are away from main traffic areas and isolated from the core campus. Access is controlled at H1 with an electronic card system, cameras, and guards. The plan is for D1 to be completely remodeled and used as a student services building only. That being said, there are still some issues that exist that the University is actively addressing. In August 2011, an employee reported that five of their DNA samples were stolen at the University. It was unclear from the police report from where these items were stolen or if these items were in a controlled location. In October 2011, a student was arrested and charged with setting off a chemical explosive at the University apartments which had an aroma of chlorine. As the University's enrollment increases and more students live on campus, the chances for these types of occurrences have increased since 2005.

The research activities that were ongoing at G1 have been further removed from main traffic areas to H1, which will help ensure safety and security of those research activities. The University Police Department has been granted more money for additional security forces, which should enable security forces to patrol a wider campus area. There are still not enough security cameras in place at the University, but this situation is gradually improving as new facilities come online.

POLICY ISSUES REGARDING HAZMAT DATA

A new issue has arisen since the first edition of this book went to press in regard to HAZMAT in general. How much access to sensitive information do you allow people who do not work for your institution? The University is located in a state that clearly specifies in their administrative codes that HAZMAT information is sensitive and should be restricted by the organization. Therefore, first responders who are not employed by the University are not authorized to review the University's data and the University has no control of other organizations' hiring practices or data policy control per state statue. The University instead allows first responders to interface directly with the University Police Department and the EHS Department to not only stay in compliance with state statute but also to better control their own data and lessen the chance of sensitive data being compromised. There is no point in providing a potential criminal with a prepared shopping list of where materials are stored and the amounts and types of specimens that the University has in use or stored at its facilities. This becomes a liability issue if the data ever falls into the wrong hands.

SUMMARY

The University has made many positive steps toward improving the safety and security of its general HAZMAT situation over the last seven years. Instead of being a typical research university with HAZMAT problems that are unresolved, the University has made a commitment to become one of the best institutions in the country for safety and security of its HAZMAT assets. There are still some shortfalls to address (e.g., more security cameras needed), but time and resources will hopefully resolve most of these issues fairly soon. The culture of the staff and faculty is changing to a more proactive stance on HAZMAT issues, but this transformation will take some time due to universities in general being slow to change. The fact that new research facilities that may contain HAZMAT are now being constructed in stand-alone buildings and isolated from the rest of the campus should indicate that the administration and faculty are adapting to new requirements and the dangers associated with new types of research. Additionally, the University has improved its response to HAZMAT issues by employing a well-trained and enthusiastic EHS staff.

Based on the findings of the security surveys, the University initiated several improvements in the way in which its research facilities are constructed and secured. The first of these improvements was initiated in 2004 with the hiring of a new environmental safety director. Staffing for the EHS Department was increased from three employees to seven employees in 2008. The department added specializations in fire safety, industrial hygiene, occupational safety, driver training, biological safety, laser/x-ray safety, and waste disposal. The EHS Department was relocated from the waste storage depot to a new building.

The new environmental safety director reviewed and modified existing safety protocols and practices. Two older science buildings at the University have had emergency response guide (ERG) chemicals broken down into different categories to aid first responders in the event of an emergency. In 2007 the environmental safety director directed the Shipping and Receiving Department to cease deliveries of compressed gas cylinders to various facilities at the campus because the delivery truck was not secure enough to deliver those types of chemicals. Now the compressed gas cylinders are directly delivered to the campus facilities by the vendors. Prior to 2004, there had been little thought given to tracking chemical HAZMAT at the University. In 2008, to comply with DHS requirements, the environmental safety staff completed a substantial inventory of existing stockpiles of chemicals located in the teaching and research laboratories. This information, once uploaded into the University's computer system, allowed authorized personnel to track laboratory chemicals, conduct audits, and respond to any crises that may occur.

Large-scale improvements have been made to campus facilities to address the heightened emphasis on HAZMAT security. In 2007, the University opened a new facility dedicated solely to research purposes that would house laboratories that were previously located in mixed-use buildings. By opening this new facility, research activities could be sequestered from other student areas and research labs could be moved from old, noncompliant areas into spaces that were safer, more secure, and compliant with current regulations (Valcik 2006). Makeshift clean rooms had been removed from an older building and relocated to this new research facility, which was specifically constructed to house clean rooms. The new facility contains chemical storage bunkers, storage for toxic gases for the clean room (which has triple redundant safety), and a monitoring system. The University has applied this new safety consciousness to a new research facility that is to be constructed and another facility for research activities (the University Environmental Safety director, interview 2008).

New security measures regarding chemical HAZMAT and general safety have been extended to existing facilities as well. Upon inspecting the X1 building, new fireproof storage cabinets were provided to store chemicals such as acids and solvents that are used in art projects. The Environmental Safety Department also recommended that X1 improve and expand its safety protocols and practices (the University Environmental Safety director, interview 2008). After 2004, room occupancies and fire codes have been taken more seriously with the addition of a fire safety specialist to the EHS Department. Some facilities have had security cameras installed. Although the security cameras are not IP addressable, they are an improvement over not having any security cameras at all.

As with any bureaucracy, some issues take time to resolve and the University is no different in this manner. As of 2008, this institution has not established a robust means whereby chemical HAZMAT can be electronically traced from moment of receipt by the shipping department to its ultimate destination on campus. There is no mechanism to track compressed gas cylinders to the various locations on campus nor has an effective bar coding system been implemented. Presently there is no campus-wide security camera plan at the University, not even for open areas such as parking lots or most building interiors (2008). Even if there was camera coverage, there is no mechanism that would enable security forces to view the camera feed in their vehicles or on foot. According to Staff Member No. 8, some cameras have been installed in B1, but these cameras were installed only after two plasma televisions had been stolen from the engineering complex. Staff Member No. 8 also stated that a recommendation was made for mounting IP addressable cameras, but efforts to purchase and install these cameras appear to have stalled.

CONCLUSIONS

Organizational changes, new guidelines, and new techniques designed to improve safety and security take time to implement. Since my research first began in 2004, the University has made great strides in improving safety and addressing homeland security issues by adopting new techniques, hiring qualified personnel, and putting forth the resources necessary to stay in compliance with the new federal guidelines. To address some of these issues, the University will require adequate resources for new construction and modifications to existing facilities and to fund new infrastructure costs (e.g., security cameras), which can be quite expensive. Therefore improvements in the HAZMAT situation at the University might take years to implement.

Although risk assessments and security surveys are often conducted by business affairs offices at most universities, such efforts can be aided, and in some cases initiated, by the institutional research office. Institutional research offices have access to personnel, student, and facilities data that can be used to analyze peak room and building usages by day and hour, the ages and dimensions of research facilities, determine departmental ownership, and define facility characteristics. Analyses such as these can assist universities in modernizing facilities, tracking and securing HAZMAT, and complying with federal and state regulations. Any attempts at improving HAZMAT security can, in turn, have a positive effect on other university initiatives, such as increasing the number and dollar value of research grants, improving student and faculty recruitment endeavors, and even accreditation efforts.

Appendix A: Representative Federal Statutes for Research Organizations with Regard to HAZMAT

APPLICABLE SUBSECTIONS OF TITLE 21 FROM THE DRUG ENFORCEMENT AGENCY

Title 21, Food and Drugs; Chapter 13, Drug Abuse Prevention and Control; Subchapter I, Control and enforcement; Part C, Registration of Manufacturers, Distributors, and Dispensers of Controlled Substances; Sec. 823, Registration requirements

(f) Research by practitioners; pharmacies; research applications; construction of Article 7 of the Convention on Psychotropic Substances

The Attorney General shall register practitioners (including pharmacies, as distinguished from pharmacists) to dispense, or conduct research with, controlled substances in schedule II, III, IV, or V, if the applicant is authorized to dispense, or conduct research with respect to, controlled substances under the laws of the State in which he practices. The Attorney General may deny an application for such registration if he determines that the issuance of such registration would be inconsistent with the public interest. In determining the public interest, the following factors shall be considered:

(1) The recommendation of the appropriate State licensing board or professional disciplinary authority.
(2) The applicant's experience in dispensing, or conducting research with respect to controlled substances.

(3) The applicant's conviction record under Federal or State laws relating to the manufacture, distribution, or dispensing of controlled substances.
(4) Compliance with applicable State, Federal, or local laws relating to controlled substances.
(5) Such other conduct which may threaten the public health and safety.

Separate registration under this part for practitioners engaging in research with controlled substances in schedule II, III, IV, or V, who are already registered under this part in another capacity, shall not be required. Registration applications by practitioners wishing to conduct research with controlled substances in schedule I shall be referred to the Secretary, who shall determine the qualifications and competency of each practitioner requesting registration, as well as the merits of the research protocol. The Secretary, in determining the merits of each research protocol, shall consult with the Attorney General as to effective procedures to adequately safeguard against diversion of such controlled substances from legitimate medical or scientific use. Registration for the purpose of bona fide research with controlled substances in schedule I by a practitioner deemed qualified by the Secretary may be denied by the Attorney General only on a ground specified in section 824(a) of this title. Article 7 of the Convention on Psychotropic Substances shall not be construed to prohibit, or impose additional restrictions upon, research involving drugs or other substances scheduled under the convention which is conducted in conformity with this subsection and other applicable provisions of this subchapter.

APPLICABLE SUBSECTIONS FROM NATIONAL INSTITUTES OF HEALTH RECOMBINANT DNA RESEARCH

Recombinant DNA Research: Actions Under the Guidelines; Notice

Any research group working with agents that are known or potential biohazards shall have an emergency plan that describes the procedures to be followed if an accident contaminates personnel or the environment.

The Principal Investigator shall ensure that everyone in the laboratory is familiar with both the potential hazards of the work and the emergency plan (see Sections IV-B-4-d and IV-B-4-e). If a research group is working with a known pathogen for which there is an effective vaccine, the vaccine should be made available to all workers. Serological monitoring, when clearly appropriate, will be provided (see Section IV-B-1-f).

The Laboratory Safety Monograph (see Appendix G-III-O) and Biosafety in Microbiological and Biomedical Laboratories (see Appendix G-III-B) describe practices, equipment, and facilities in detail. The institution shall establish and maintain a health surveillance program for personnel engaged in animal research involving viable recombinant DNA-containing microorganisms that require Biosafety Level (BL) 3 or greater containment in the laboratory.

Appendix Q-II-A-1-a-(1). The containment area shall be locked.

Appendix Q-II-A-1-a-(2). Access to the containment area shall be limited or restricted when experimental animals are being held.

Appendix Q-II-A-1-a-(3). The containment area shall be patrolled or monitored at frequent intervals.

Appendix Q-II-A-1-b. Other (BL1-N)

Appendix Q-II-A-1-b-(1). All genetically engineered neonates shall be permanently marked within 72 hours after birth, if their size permits. If their size does not permit marking, their containers should be marked. In addition, transgenic animals should contain distinct and biochemically assayable DNA sequences that allow identification of transgenic animals from among non-transgenic animals.

Appendix Q-II-B-1-g-(8). A biosafety manual shall be prepared or adopted. Personnel shall be advised of special hazards and required to read and follow instructions on practices and procedures.

Appendix Q-II-C-1-b-(4). Special safety testing, decontamination procedures, and Institutional Biosafety Committee approval shall be required to transfer agents or tissue/organ specimens from a BL3-N animal facility to a facility with a lower containment classification.

Appendix Q-II-C-1-c-(1). When the animal research requires special provisions for entry (e.g., vaccination), a warning sign incorporating the universal biosafety symbol shall be posted on all access doors to the animal work area. The sign shall indicate: (i) The agent, (ii) the animal species, (iii) the name and telephone number of the Animal Facility Director or other responsible individual, and (iv) any special requirements for entering the laboratory.

Appendix Q-II-C-1-d-(3). Appropriate respiratory protection shall be worn in rooms containing experimental animals.

Appendix Q-II-C-1-e. Records (BL3-N)

Appendix B: Questions Asked of Faculty about the Evolution of the University

Hello, I am Nicolas Valcik and I am a doctoral student. I am currently working on my dissertation that is researching organizational change and decision making in research activities at universities that transition from rural to urban environments. I would like to interview you to gain insight into how the University has changed from its beginning as a research university in a rural setting to a university of over 14,000 students in a suburban environment.

1. What year did you come to the University?
2. What position did you start at in the organization?
3. Are you still working at the University? If not, what year did you leave or retire?
4. What is your current or last position with the University?
5. When you came to the University, were the decisions in the organization made primarily by the individual researcher or were decisions made primarily by the administration?
6. At which level were these decisions (or are these decisions) made?
7. How has the work environment changed from when you originally started working at the university to the present?
8. Was more focus given to the Natural Science and Mathematics (NS&M) faculty staff originally when you were hired compared to the present?
9. Were research facilities adequate when you originally arrived at the university for the school of NS&M?
10. Have the facilities for NS&M research been adequate during your tenure?
11. If not, when did the facilities cease to be adequate for research operations and in what areas were facilities deficient?
12. If not adequate, has the research facility situation improved since the deficiency occurred?

13. If the situation has not improved, to what factors do you attribute the shortfall of adequate facilities to perform research (e.g., lack of budget, lack of administration focus on NS&M, etc.)?
14. If a shortfall has occurred, what have been the consequences from that facility shortfall?
15. Have resources in the form of personnel or facilities expanded adequately with the expansion of research activities throughout the years?
16. How has the transition of the University from a rural to a suburban environment affected research activities and organizational decision making?
17. What have been the benefits on research activities of having more students at the University?
18. What have been the determents on research activities of having more students at the University?
19. How has the increase in students affected organizational decision making with regard to research activities?

I would like to thank you for your time and effort during this interview.

Appendix C: Interview Questions Asked of Staff Members on Current Hazmat Policies, Practices, and Procedures

1. What is your current position at the University?
2. When did you start working for the University?
3. What is your current responsibility with regard to compliance with federal and state HAZMAT regulations with regard to researchers and the contracts received?
4. Are the facilities up to par for research contracts and grants received?
5. Do we follow CDC/NIH standards for lab construction on anything related to biologic research?
6. What laboratory protocols are currently standard and at use in the University?
7. What biosafety level is the Research One building currently configured?
8. Does the University currently address the use of international students on contracts and grants with regard to the Patriot Act?
9. What is the procedure for ensuring compliance of animal and human experimentation to federal guidelines?
10. Who ensures compliance at the University with regard to this?
11. What mechanism determines the specifications that laboratories are constructed at?
12. What are the policies for access to Research One and transportation of animals across campus?

Appendix D: Interview Questions Asked of Former Assistant Director of Environmental Health and Safety at University of Texas Southwestern (UTSW) Medical Center on HAZMAT Policies, Practices, and Procedures

1. What is your current responsibility with regard to compliance with federal and state HAZMAT regulations with regard to researchers and the contracts received?
2. What was your position at UTSW Medical Center?
3. When did you start working at UTSW Medical Center?
4. What laboratory protocols did UTSW Medical Center use for biological policies and procedures?
5. How did UTSW Medical Center insure that policies and procedures were followed by researchers?
6. Did UTSW Medical Center follow NIH/CDC guidelines in the construction of laboratories?
7. Did UTSW Medical Center follow the CLIA 1988 guidelines for laboratories that dealt with patients, blood, or tissue samples?
8. Do you have a current set of UTSW Medical Center procedures?
9. Do other UT-system campuses follow these protocols?

10. Did UTSW Medical Center keep a centralized list of biological elements in a centralized database for inventory control or first responders?
11. Did UTSW Medical Center address the use of international students on contracts and grants with regard to the Patriot Act?
12. What is the procedure for ensuring compliance of animal and human experimentation to federal guidelines for UTSW Medical Center?
13. Who ensures compliance at UTSW Medical Center with regard to this?
14. What mechanism determines the specifications that laboratories are constructed at for UTSW Medical Center?
15. What are the policies for access to the animal care facilities and transportation of animals across the UTSW Medical Center campus?

Appendix E: Interview Questions to Faculty Senate and Safety Committee

1. What year did you come to the University?
2. How long have you been on the Faculty Senate?
3. What positions in the Faculty Senate have you held?
4. Do you currently deal with HAZMAT for research purposes?
5. What is the history of the Safety Committee?
6. What issues is the committee charged with?
7. Does the University have HAZMAT policies and procedures?
8. If so, what year were the policies and procedures implemented?
9. If no policies and procedures exist, are there plans to formulate such policies and procedures?
10. Is the administration such as the police and Environmental Safety aware of the policies and procedures?
11. If not, why have the policies and procedures not been forwarded on to those departments?
12. What is your knowledge on current practices and procedures for research with regard to HAZMAT?
13. Are you aware of EPA guidelines regarding hazardous waste?
14. Do you feel faculty and administration have an equal concern with regard to HAZMAT policies and procedures?
15. If not, why?
16. What is the art studio's HAZMAT condition?
17. What is the Founder's building HAZMAT condition?
18. What is the engineering computer science HAZMAT condition?
19. Do you feel that research faculty should be accountable for HAZMAT concerns that arise due to their experiments?
20. What role do you feel that administration should play with regard to HAZMAT policies and procedures?

Thank you for your time during this interview.

Appendix F: Interview with Former Director at Research Think Tank

1. What was your former position at the Research Think Tank?
2. When did you start working for the Research Think Tank?
3. When did you leave the Research Think Tank?
4. What were your responsibilities at the Research Think Tank?
5. What was the type of research being performed at the Research Think Tank?
6. Was this type of research related to national defense, or nuclear, biological, and chemical (NBC) warfare?
7. Were there centralized HAZMAT policies for the Research Think Tank?
8. What was the purpose of the Research Think Tank?
9. Why was it founded?
10. Why were the refrigerators put into the hallway?
11. When did this occur?
12. How did education offerings impact the Research Think Tank?
13. How was biological HAZMAT handled by the researchers?
14. Was the waste produced from experiments the responsibility of the Research Think Tank or the researcher?
15. Where were the animals for research initially housed?

Appendix G: National Science Foundation Rankings

National Science Foundation Rankings – B-33. Federally Financed R&D Expenditures at Universities and Colleges, Ranked by Fiscal Year 2002 Federally Financed R&D Expenditures: Fiscal Years 1995 to 2002

Institution and Ranking	2002	Institution and Ranking	2002
1 Johns Hopkins U., The[1]	1,022,510	312 DE State U.	2,819
2 U. WA	487,059	313 CUNY Herbert H. Lehman C.	2,816
3 U. MI all campuses	444,255	314 CA State Polytechnic U. Pomona	2,806
4 Stanford U.	426,620	315 CA State U. Dominguez Hills	2,783
5 U. PA	397,587	316 U. Redlands	2,674
6 U. CA Los Angeles	366,762	317 Maharishi U. of Management	2,618
7 U. CA San Diego	359,383	318 U. NE Omaha	2,613
8 Columbia U. in the City of New York	356,749	319 Air Force Institute of Technology	2,609
9 U. WI Madison	345,003	320 CUNY Brooklyn C.	2,538
10 U. CO all campuses	340,466	321 Lincoln U. (Jefferson City, MO)	2,513
11 Harvard U.	336,607	322 TX Christian U.	2,492
12 MA Institute of Technology[2]	330,409	323 Sam Houston State U.	2,386
13 U. CA San Francisco	327,393	324 C. Charleston	2,364
14 U. Pittsburgh all campuses	306,913	325 U.S. Naval Academy	2,324
15 Washington U. St. Louis	303,441	326 Duquesne U.	2,232
16 U. MN all campuses	295,301	327 Savannah State U.	2,206
17 PA State U. all campuses	284,706	328 Bowie State U.	2,200
18 Yale U.	274,304	329 U. AR Little Rock	2,088 e
19 Cornell U. all campuses[2]	270,578	330 Lamar U.	2,080
20 U. Southern CA	266,645	331 St. Joseph's U.	2,069
21 Duke U.	261,356	332 Ithaca C.	2,060
22 Baylor C. of Medicine	259,475	333 Winston-Salem State U.	2,022
23 U. NC Chapel Hill	254,571	334 SC State U.	1,998
24 U. TX Austin	219,158	335 Fordham U.	1,990
25 U. CA Berkeley[2]	217,297	336 Grambling State U.	1,977

Institution and Ranking	2002	Institution and Ranking	2002
26 U. AL Birmingham, The	216,221	337 DePaul U.	1,971
27 U. IL Urbana-Champaign	214,323	338 Nova Southeastern U.	1,959
28 U. AZ	211,772	339 TX A&M U. Kingsville	1,956
29 CA Institute of Technology[2]	199,944	340 NM Highlands U.	1,953
30 U. Rochester[2]	195,298	341 CA State U. Hayward	1,930
31 U. MD College Park	194,095	342 Gallaudet U.	1,880
32 Emory U.	186,083	343 Northeastern OH Universities C. of Medicine	1,871
33 U. Chicago[2]	183,830	344 Morehouse C.	1,843i
34 Case Western Reserve U.	181,888	345 New England C. of Optometry	1,792
35 U. IA	180,743	346 TN Technological U.	1,695
36 Northwestern U.	178,607	347 Morgan State U.	1,674
37 OH State U. all campuses	177,883	348 Western WA U.	1,651
38 U. CA Davis	176,644	349 CUNY York C.	1,650
39 Vanderbilt U.	172,858	350 Albany State U.	1,64
40 Boston U.	171,438	351 Ball State U.	1,590
41 U. FL	167,108	352 IL State U.	1,574
42 GA Institute of Technology all campuses	165,680	353 U. New Haven	1,492 e
43 TX A&M U. all campuses	163,488	354 Central WA U.	1,468
44 U. TX Southwestern Medical Ctr. Dallas	155,258	355 Mt. Holyoke C.	1,449
45 U. VA all campuses	152,358	356 U. San Diego	1,425
46 U. Cincinnati all campuses	150,166	357 Southern IL U. Edwardsville	1,413
47 NY U.	149,995	358 CA State U. San Bernardino	1,39
48 U. IL Chicago	143,183	359 U. North FL	1,397
49 U. UT	142,625	360 CA State U. Fresno	1,392
50 Carnegie-Mellon U.[2]	137,967	361 Williams C.	1,389
51 IN U. all campuses	132,759	362 U. TX-Pan American	1,387
52 OR Health Sciences U.	130,231	363 James Madison U.	1,372
53 SUNY Buffalo all campuses	128,842	364 U. Richmond	1,347
54 Mt. Sinai School of Medicine	125,979	365 TX Woman's U.	1,321
55 MI State U.	122,595	366 Harvey Mudd C.	1,316 i
56 U. Miami	121,171	367 CUNY C. Staten Island	1,315
57 U. TX M. D. Anderson Cancer Ctr.	117,633	368 Occidental C.	1,303
58 U. MD Baltimore	117,017	369 CA State U. Monterey Bay	1,285 e

Institution and Ranking	2002	Institution and Ranking	2002
59 U. CA Irvine	115,548	370 St. Cloud State U.	1,260 i
60 Yeshiva U.	114,268	371 AL State U.	1,249
61 CO State U.	112,650	372 Bucknell U.	1,240
62 U. HI Manoa	110,882	373 Chicago State U.	1,236
63 SUNY Stony Brook all campuses	108,122	374 Union C. (Schenectady, NY)	1,217
64 Purdue U. all campuses	107,477	375 U. WI La Crosse	1,196
65 U. NM all campuses	104,252	376 SUNY C. Old Westbury	1,193
66 U. KY all campuses	100,426	377 Long Island U. all campuses	1,162
67 U. TX Houston Health Science Ctr.	98,676	378 Middlebury C.	1,150
68 LA State U. all campuses	97,928	379 U. Central AR	1,140
69 Princeton U.[2]	97,724	380 U.S. Coast Guard Academy	1,128
70 Wayne State U.	95,910	381 Swarthmore C.	1,107
71 U. MA Worcester	93,992	382 Hampshire C.	1,095
72 U. CT all campuses	93,326	383 New School U.	1,083 e
73 Wake Forest U.	91,738	384 U. Northern IA	1,063
74 OR State U.	91,683	385 St. John's U. (Jamaica, NY)	1,060 i
75 Rutgers The State U. NJ all campuses	91,205	386 Elizabeth City State U.	1,058
76 U. of Medicine and Dentistry NJ	90,235	387 Macalester C.	1,057
77 U. TN system	88,344	388 Stephen F. Austin State U.	1,054
78 Dartmouth C.	87,255	389 Towson U.	1,037
79 Georgetown U.	87,087	390 Hamilton C.	1,035
80 U. South FL	84,108	391 SUNY C. of Optometry	1,032
81 U. TX Health Science Ctr. San Antonio	83,761	392 MS Valley State U.	1,030
82 VA Polytechnic Institute & State U.	82,976	393 Barnard C.	1,021
83 U. KS all campuses	82,663	394 Furman U.	1,003
84 UT State U.	79,393	395 U. WI Stout	1,003
85 Thomas Jefferson U.	79,217	396 Eastern WA U.	986
86 Woods Hole Oceanographic Institution	78,458	397 Bates C.	977
87 U. CA Santa Barbara	78,370	398 NC Central U.	971 i
88 U. TX Medical Branch Galveston	78,100	399 CUNY Medgar Evers C.	970
89 U. GA	78,086	400 Benedict C.	966
90 U. MO-Columbia	77,742	401 Pittsburg State U.	963
91 MS State U.	77,521	402 Carleton C.	959

Institution and Ranking	2002	Institution and Ranking	2002
92 Medical C. WI	76,241	403 Amherst C.	955
93 Medical U. SC	75,803	404 Bradley U.	953
94 NC State U.	75,204	405 Alfred U. all campuses	946
95 Tufts U.	73,236	406 PA C. of Optometry	928
96 IA State U.[2]	71,419	407 Northeastern State U.	919
97 FL State U.	70,456	408 West Chester U. PA	918
98 NM State U. all campuses	70,314	409 Pomona C.	917
99 U. OK all campuses	69,388	410 Southern U. New Orleans	902
100 Brown U.	68,215	411 SUNY C. Oswego	901
101 Rockefeller U.	67,559	412 Kirksville C. of Osteopathic Medicine	895
102 U. AK Fairbanks all campuses	66,169	413 GA Southern U.	882
103 VA Commonwealth U.	65,028	414 Haverford C.	874
104 Tulane U.	64,824	415 Hope C.	853
105 AZ State U. main campus	64,407	416 U. Northern CO	840
106 George Washington U.	58,729	417 CUNY Graduate Ctr.	832
107 U. VT	58,280	418 Baylor U.	830
108 WA State U.	56,360	419 Youngstown State U.	800
109 U. MA Amherst	54,770	420 Bethune Cookman C.	794
110 U. SC all campuses	53,403	421 Coastal Carolina U.	787
111 U. NE Lincoln	51,405	422 Eastern MI U.	777
112 U. NH	50,829	423 Loyola C.	770
113 U. MS all campuses	50,092 i	424 SUNY C. Plattsburgh	760
114 WV U.	49,394	425 Bryn Mawr C.	751
115 U. DE	48,183	426 Rose-Hulman Institute of Technology	744
116 Clemson U.	47,174	427 Murray State U.	739
117 U. RI	45,453	428 U. of the Pacific	730
118 Rush U.	44,756	429 OR Institute of Technology	727
119 Temple U.	44,577	430 Lincoln U. (Lincoln University, PA)	718
120 Naval Postgraduate School	44,379 i	431 Santa Clara U.	706 i
121 KS State U.	43,988	432 Drake U.	692
122 Auburn U. all campuses	42,432	433 Seton Hall U.	692
123 SUNY Albany	40,497	434 Dickinson C.	685
124 U. Dayton	40,191	435 St. Olaf C.	683
125 MT State U.-Bozeman	39,986	436 U. AK Southeast	673 e
126 U. NV, Reno	39,958	437 Bowdoin C.	658
127 Rice U.	39,739	438 Augsburg C.	655

Institution and Ranking	2002	Institution and Ranking	2002
128 Charles R. Drew U. of Medicine & Science	39,259	439 Milwaukee School of Engineering	651
129 U. Louisville	38,978	440 U. WI Stevens Point	639
130 U. OR	37,177	441 Kettering U.	638
131 U. AL Huntsville, The	35,660	442 Claremont Graduate U.	621
132 Howard U.	33,949	443 CA State U. Chico	617
133 U. ID	33,465	444 U. WI Green Bay	601
134 U. of Notre Dame	33,285	445 U. Houston-Downtown	596
135 U. CA Santa Cruz	32,901	446 Midwestern U.	589
136 Syracuse U. all campuses	32,704	447 Colgate U.	586
137 U. CA Riverside	32,305	448 Bennett C.	583 i
138 FL International U.	32,057	449 Reed C.	583
139 George Mason U.	31,890	450 SUNY C. Buffalo	583
140 U. Houston	31,455	451 U. WI Eau Claire	580
141 St. Louis U. all campuses	31,215	452 Albany C. of Pharmacy	572
142 OK State U. all campuses	31,100	453 U. WI Parkside	563
143 Drexel U.	30,232	454 Allegheny C.	562
144 U. Central FL	29,929	455 U. WI Superior	552
145 U. MD Baltimore County	29,376	456 Appalachian State U.	543
146 Brandeis U.	29,006	457 Southern OR U.	542
147 U. NE Medical Ctr.	28,602	458 U. Southern CO	541
148 TX Tech U.	28,202	459 U. Hartford	537
149 Loyola U. (Chicago, IL)	27,678	460 VA Union U.	537 i
150 U. AR main campus	27,588	461 Southeastern LA U.	525
151 U. AR for Medical Sciences	27,491	462 CA State U. Bakersfield	523 i
152 U. AL, The	26,900	463 Lafayette C.	519
153 MCP Hahnemann U.	26,619	464 Smith C.	515
154 Rensselaer Polytechnic Institute	26,490	465 U. of Health Sciences	513
155 NJ Institute of Technology	25,641	466 John Carroll U.	489
156 SUNY Health Science Ctr. Brooklyn	25,495 i	467 SUNY C. Geneseo	481
157 San Diego State U.	25,223	468 Indiana U. PA all campuses	476
158 U. ND all campuses	25,004	469 C. NJ, The	456
159 Northeastern U.	24,295	470 Oakwood C.	447 i
160 U. PR Medical Sciences Campus	23,931	471 Chapman U.	442
161 U. ME	23,732	472 Evergreen State C.	441
162 Medical C. GA	23,317	473 Franklin and Marshall C.	441
163 Boston C.	22,827	474 U. NC Asheville	412

Institution and Ranking	2002	Institution and Ranking	2002
164 Desert Research Institute	22,760	475 McNeese State U.	409
165 U. MT, The	22,612	476 Tougaloo C.	407 i
166 TX A&M U. System Health Science Ctr.	22,417	477 Oberlin C.	406
167 U. Southern MS	22,398	478 Grand Valley State U.	405
168 U. NV, Las Vegas	22,367	479 Hofstra U.	402 e
169 Morehouse School of Medicine	21,919	480 National U. of Health Sciences	402
170 Jackson State U.	21,721	481 State U. West GA	400
171 FL A&M U.	21,592	482 Dillard U.	393
172 ND State U. all campuses	21,414	483 SUNY C. Brockport	392 i
173 Meharry Medical C.	20,453	484 Philadelphia C. of Osteopathic Medicine	387
174 Loma Linda U.	20,114	485 Grinnell C.	385
175 U. WY	20,017 e	486 Montclair State U.	383
176 NY Medical C.	19,318	487 U. Detroit Mercy	369
177 GA State U.	18,967	488 Pace U. all campuses	362
178 NM Institute of Mining and Technology	18,864	489 Spalding U.	359
179 C. of William and Mary all campuses	18,007	490 Radford U.	357
180 MI Technological U.	17,973	491 U. WI Oshkosh	356
181 Catholic U. America	17,925	492 CUNY Baruch C.	347
182 OH U. all campuses	17,677	493 Morris Brown C.	343
183 Old Dominion U.	17,170	494 Calvin C.	338
184 U. TX El Paso	16,457	495 C. of the Holy Cross	329
185 Wright State U. all campuses	16,070	496 SUNY C. Cortland	316
186 U. South AL	15,922	497 AR State U. main campus	312
187 San Jose State U.	15,777	498 Fayetteville State U.	311
188 U. MO-Rolla	15,749	499 U. of the Sciences in Philadelphia	311
189 FL Atlantic U.	15,566	500 Colby C.	306
190 Clark Atlanta U.	15,325	501 U. Portland	285
191 U. MA Lowell	14,932	502 C. Wooster	284
192 CUNY Hunter C.	14,573	503 IN State U. all campuses	284
193 U. Memphis, The	14,072	504 Central State U.	283 i
194 Lehigh U.	13,519	505 Western Carolina U.	280
195 SUNY Upstate Medical U.	13,404	506 Lewis and Clark C.	272
196 CUNY City C.	13,319	507 Augustana C. (Sioux Falls, SD)	265
197 CO School of Mines	13,230	508 St. Mary's C. MD	262

Institution and Ranking	2002	Institution and Ranking	2002
198 U. MD Ctr. for Environmental Science	13,080	509 Valparaiso U.	262
199 Eastern VA Medical School	13,077	510 Manhattan C.	257
200 U. LA Lafayette	12,925	511 Trinity U.	250
201 Brigham Young U. all campuses	12,900	512 U. Central OK	250
202 Medical C. OH	12,006	513 Eastern OR U.	243
203 NC Agricultural and Technical State U.	11,953	514 Wilberforce U.	236 i
204 Portland State U.	11,914	515 Pitzer C.	227
205 U. TX Dallas	11,624	516 Skidmore C.	227
206 U. WI Milwaukee	11,461	517 Eastern KY U.	225
207 IL Institute of Technology	11,277	518 Ft. Lewis C.	221
208 TN State U.	11,218	519 Kennesaw State U.	221
209 Northern AZ U.	10,873	520 Rider U.	218
210 U. MO-Kansas City	10,795	521 Alliant International U.	214
211 Southern IL U. Carbondale	10,789	522 Western U. of Health Sciences	214
212 AL A&M U.	10,710	523 Middle TN State U.	211
213 U. of the VI	10,620	524 Whitman C.	211
214 U. MD Biotechnology Institute	10,603	525 U. WI Whitewater	210
215 Uniformed Services U. of the Health Sciences	10,294	526 Claflin C.	203
216 Hampton U.	10,058	527 OH Wesleyan U.	203
217 U. NC Wilmington	9,950	528 CUNY John Jay C. of Criminal Justice	201
218 Western MI U.	9,913	529 U. IL Springfield	200
219 U. Akron all campuses	9,537	530 Bridgewater State C.	196
220 Marshall U. Graduate C.	9,502	531 Lawrence Technological U.	191
221 Kent State U. all campuses	9,427	532 Central MI U.	190
222 SD State U.	9,115	533 NJ City U.	184
223 U. Toledo	8,996	534 Denison U.	183
224 SUNY Binghamton	8,959	535 MN State U. Mankato	181
225 Albany Medical C.	8,901	536 Trinity C. (Hartford, CT)	181
226 U. PR Mayaguez Campus	8,865	537 Saginaw Valley State U.	174
227 Mercer U. all campuses	8,607	538 Abilene Christian U.	173
228 U. HI Hilo	8,499	539 Providence C.	173
229 U. Denver	8,386	540 Iona C.	167
230 U.S. Air Force Academy	8,163	541 Des Moines U. Osteopathic Medical Ctr.	164 e

Institution and Ranking	2002	Institution and Ranking	2002
231 Stevens Institute of Technology	8,094 e	542 Sonoma State U.	161
232 Miles C.	7,943	543 U. of St. Thomas (St. Paul, MN)	160
233 Southern U. A&M C. all campuses	7,903 i	544 Pacific U.	157
234 Rochester Institute of Technology	7,868	545 U. Southern ME	157 i
235 U. TX Arlington	7,848	546 St. Mary's U. (San Antonio, TX)	152
236 U. North TX Health Science Ctr. Ft. Worth	7,770	547 SUNY Purchase C.	151
237 CA State U. Los Angeles	7,728	548 Southern CT State U.	144
238 Prairie View A&M U.	7,712	549 CO C.	143
239 Tuskegee U.	7,703	550 AR Tech U.	130 e
240 U. Houston-Clear Lake	7,460	551 Lake Forest C.	126
241 U. MA Dartmouth	7,142	552 Andrews U.	123
242 U. TX San Antonio	7,111	553 Simmons C.	118
243 SD School of Mines and Technology	6,950	554 Adelphi U.	114 i
244 VA State U.	6,895	555 U. NE Kearney	108
245 U. Tulsa	6,806	556 Fairfield U.	106
246 CA State U. Northridge	6,786	557 Salem International U.	106 e
247 Wichita State U.	6,687	558 Regis U.	105
248 Creighton U.	6,594	559 SUNY C. Fredonia	104
249 Universidad Central Del Caribe	6,470	560 Central CT State U.	102
250 KY State U.	6,036	561 Juniata C.	101
251 East Carolina U.	6,030	562 Shaw U.	100 i
252 Southern Methodist U.	6,026	563 U. Findlay, The	100
253 Xavier U. LA	6,019	564 Fairleigh Dickinson U. all campuses	98
254 U. North TX	6,015	565 U. WI River Falls	89
255 CA State Polytechnic U. San Luis Obispo	5,968	566 Willamette U.	87
256 Teachers C. Columbia U.	5,939	567 NY Institute of Technology all campuses	84
257 U. NC Charlotte	5,850	568 Southwest MO State U.	79
258 U. SD	5,830	569 Wilkes U.	79
259 Alcorn State U.	5,809	570 MA C. of Pharmacy & Health Sciences	77

Institution and Ranking	2002	Institution and Ranking	2002
260 Finch U. of Health Sciences/Chicago Medical School	5,638	571 Sul Ross State U.	77
261 CUNY Queens C.	5,516	572 Monmouth U.	73
262 U. West FL	5,315	573 Truman State U.	68
263 Tarleton State U.	5,216	574 Ferris State U.	62 e
264 Worcester Polytechnic Institute	5,013	575 Knox C.	60
265 Western IL U.	4,982	576 Nicholls State U.	59
266 Northern IL U.	4,980	577 New C. FL	55
267 U. PR Rio Piedras Campus	4,971	578 Pontifical Catholic U. PR, The	54
268 CA State U. Long Beach	4,944	579 Point Loma Nazarene C.	53
269 Polytechnic U.	4,918	580 Northeastern IL U.	47
270 Oakland U.	4,903	581 AK Pacific U.	43
271 Norfolk State U.	4,902	582 Widener U. all campuses	42 i
272 U. MA Boston	4,837	583 TX A&M U. Commerce	37
273 U. MO-St. Louis	4,755	584 Southeast MO State U.	30
274 ID State U.	4,738	585 U. Dallas	30
275 Cleveland State U.	4,649	586 Northern KY U.	26
276 SUNY C. of Environmental Science & Forestry	4,630	587 Wentworth Institute of Technology	24
277 Villanova U.	4,551 e	588 Western State C. CO	17
278 CA State U. Fullerton	4,434	589 San Francisco State U.	13
279 Clarkson U.	4,286e	590 Jarvis Christian C.	10
280 U. AK Anchorage all campuses	4,244 e	591 VA Military Institute	10
281 U. GU	4,183	592 Antioch U. all campuses	0
282 Embry-Riddle Aeronautical U.	4,174	593 Azusa Pacific U.	0
283 Boise State U.	4,173	594 Biola U.	0
284 Southwest TX State U.	4,144	595 Bluefield State C.	0
285 Miami U. all campuses	4,021	596 CA Institute of Integral Studies	0
286 Langston U.	3,975	597 CT C.	0 i
287 TX Southern U.	3,910	598 Cooper Union	0
288 East TN State U.	3,908	599 Coppin State C.	0
289 Wesleyan U.	3,734	600 East Stroudsburg U. PA	0
290 Ponce School of Medicine	3,720	601 Forest Institute of Professional Psychology	0
291 U. Scranton	3,715	602 Golden Gate U.	0

Institution and Ranking	2002	Institution and Ranking	2002
292 FL Institute of Technology	3,646	603 Institute of Textile Technology	0
293 Marquette U.	3,565	604 Johnson C. Smith U.	0
294 U. LA Monroe, The	3,540 i	605 La Salle U.	0
295 Western KY U.	3,493	606 Le Moyne-Owen C.	0
296 LA Tech U.	3,487	607 Pepperdine U.	0
297 American U.	3,467	608 Philander Smith C.	0
298 Wellesley C.	3,446e	609 Rust C.	0
299 Clark U.	3,418	610 Selma U.	0
300 Humboldt State U.	3,354 i	611 Talladega C.	0
301 U. NC Greensboro	3,340	612 Union Institute, The	0
302 West TX A&M U.	3,245	613 U. MD Eastern Shore	0
303 TX A&M U. Corpus Christi	3,127	614 U. MO system administration	0
304 Fisk U.	3,030	615 WV School of Osteopathic Medicine	0
305 U. AR Pine Bluff	3,016	616 WV State C.	0 i
306 Ft. Valley State U.	2,985	617 Wiley C.	0
307 MT Tech of The U. MT	2,977		
308 Spelman C.	2,968		
309 U. DC	2,963		
310 Institute of Paper Science and Technology	2,947e		
311 Bowling Green State U. all campuses	2,829		

Note: Dollars in thousands. Because of rounding, detail may not add to totals. e = estimated. i = imputed. — = not available.

[1] Johns Hopkins University includes the Applied Physics Laboratory, with $540 million in federally financed R&D expenditures.

[2] These data do not include R&D expenditures at university-associated federally funded research and development centers.

Bibliography

Associated Press. 2001. "Polluted campuses: EPA fines more colleges." CNNfyi.com: Education News, March 9.

Associated Press. 2004a. "A&M apartment blast prompts new inspections." HoustonChronicle.com. Retrieved on October 11, 2004.

Associated Press. 2004b. "St. Ed's, UT, Texas State hit in string of computer thefts." Retrieved on October 25, 2004, from http://www.kvue.com/news/top/stories/102504kvuecomputerthefts-jw.228b4971.html.

Associated Press. 2008. "UT lab safety expert says he was fired for finding problems." *The Fort Worth Star-Telegram*, May 31. Retrieved on May 31, 2008, from http://www.star-telegram.com/448/story/673569.html.

"A Balancing Act? Openness and Security on Campus." 2004. *Syllabus* 17(9).

Bartlett, Thomas, and Wilson, Robin. 2010. "The fatal meeting." *The Chronicle of Higher Education* 56(24).

Becker, Howard S., Geer, Blanche, Riesman, David, and Weiss, Robert S. 1968. *Institutions and the person: Papers presented to Everett C. Hughes*. Chicago, IL: Aldine.

Bellamy, Patrick. 2004. "False prophet: The Aum cult of terror." Court TV Crime Library, http://www.crimelibrary.com/terrorists_spies/terrorists/prophet/1.html?sect=22.

Benke, Richard. 2004. "2 former Los Alamos lab workers indicted." Associated Press, May 26.

Berger, Eric. 2005. "Biosafety lab may not be as open as promised." *The Houston Chronicle*, May 9, http://www.chron.com/cs/CDA/printstory.mpl/front/3172872.

Binghamton University, State University of New York. 2004. "Division of administration, organizational chart," http://www.binghamton.edu/.

Binghamton University, State University of New York. 2012. Environmental Health and Safety. Retrieved on January 11, 2012, from http://www.binghamton.edu/.

Brainard, Jeffery. 2005. "Federal agencies issue final rules on safeguarding academic research on dangerous microbes." *The Chronicle of Higher Education*, March 23.

Campbell, Donald T., and Stanley, Julian C. 1963. *Experimental and quasi-experimental designs for research*. Chicago, IL: Rand-McNally.

Campbell, Kenneth D. 2002. "MIT panel urges off-campus sites for classified research; reaffirms openness of MIT campus." *MIT News*, June 12.

Centers for Disease Control and Prevention (CDC)/National Institutes of Health (NIH). 2004. *Biosafety in Microbiological and Biomedical Laboratories*.

Cohen, Robert. 2003. "Campus fire safety right to know bill gets new life." *Star-Ledger*, July 10, http://www.freerepublic.com/focus/f-news/927882/posts.

Corbett, Molly. 1999. "The University of North Carolina: Responding to the aftermath of Hurricane Floyd." Presentation October 1999 to the University of North Carolina Board of Governors, http://www.ga.unc.edu/BOG/BOG_REPORTS/BR1999_10.pdf.

Cyert, R. M. 1978. "The management of universities of constant or decreasing size." *Public Administration Review* 38: 344–349.

Daft, Richard L. 1995. *Organization theory and design*, 5th ed. New York, NY: West Publishing.

Denzin, Norman K. 1978. *Sociological methods: A sourcebook*. New York, NY: McGraw-Hill.

Department of Energy. 2001. "Decommissioning the Brooklyn National Laboratory Building 830 Gamma Irradiation Facility." Report Number: BNL—52637: YN0100000, August 13.

Department of Health and Human Services. 2004. "Summary report on select agent security at universities." Office of Inspector General, A-04-04-02000, March 25.
Department of Health and Human Services. 2006. "Summary report on universities' compliance with select agent regulations," Office of Inspector General, A-04-05-02006, June 30.
Department of Homeland Security. 2003. "Combating Terrorism Technology Support Office (CTTSO) Technical Support Working Group (TSWG) Broad Agency announcement," TSWG BAA 03-Q-4070.
Department of Homeland Security. 2007. *Department of Homeland Security Appropriations Act 2007: Section 550, DHS-2006-0073, RIN 1601-AA41, 6 CFR Part 27—Chemical Facility Anti Terrorism Standards.* Retrieved on January 16, 2009. http://www.dhs.gov/xlibrary/assets/IP_ChemicalFacilitySecurity.pdf.
Department of Homeland Security. 2009. "DHS issues record of decision on proposed bio- and agro-defense facility." Retrieved on January 13, 2011, from http://www.dhs.gov/xnews/releases/pr_1232132671186.shtm.
Department of Homeland Security. 2011. "History." Retrieved on December 21, 2011, from http://www.dhs.gov/xbout/history/.
Downs, Anthony. 1967. *Inside bureaucracy*. Boston, MA: Little Brown.
Drew, Clifford J., and Hardman, Michael L. 1985. *Designing and conducting behavioral research*. New York, NY: Pergamon Press.
Drug Enforcement Agency. 2004. "DEA genealogy." http://www.usdoj.gov/dea/history.htm.
Eiserink, Martin, and Malakoff, David, 2003. "The Trials of Thomas Butler." *Science Magazine*, December 19, http://www.freerepublic.com/focus/f-news/1044065/posts.
Elias, Paul. 2004. "'Hot lab' boom raises concerns." *Dallas Morning News*, May 2.
Environmental Protection Agency. 2004a. Colleges and Universities, Sectors, EPA. Presented by Gerald Carney and Marcia Moncrieff. Eighth Annual Texas Law Conference at the University of North Texas in Denton, Texas, March 1–2.
Environmental Protection Agency. 2004b. "Our history." http://www.epa.gov/history/.
Federal Bureau of Investigation (FBI). 1970. "Kent State shooting," file number 98-46479, http://foia.fbi.gov/kentstat.htm.
Federal Emergency Management Agency (FEMA). 2003. Pre-Disaster Mitigation Disaster Resistant University Grants.
Field, Kelly. 2004. "Security at university labs was lax, federal report says." *Chronicle of Higher Education*, April 22.
Field, Kelly. 2005. "Biosafety committees come under scrutiny." *Chronicle of Higher Education*, April 29.
Fischman, Josh. 2012. "Science and security clash on bird-flu papers." *Chronicle of Higher Education* 58(18).
Fleming, Diano O., and Hunt, Debra L. (eds.). 2000. *Biological safety: Principles and practices*. Washington, DC: ASM Press.
Fogg, Piper. 2004. "Professor at U. of Louisiana at Lafayette is suspended after threatening to kill students." *Chronicle of Higher Education*, October 15.
Franconere, Vincent, 2004. Director of Environmental Health and Safety at SUNY – Albany.
Georgia Tech. 2012. "Environmental health and safety." Retrieved on January 11, 2012, from http://www.safety.gatech.edu/.
Glaser, Barney G., and Strauss, Anselm L. 1967. *The discovery of grounded theory: Strategies for qualitative research*. Chicago, IL: Aldine.
Graduate Research Center of the Southwest. 1961. *Annual report*.
Grayson, Katherine. 2005. "Campus technology innovations." *Campus Technology* 18(12).

Gross, Bertram M. 1968. *Organizations and their managing*. New York, NY: The Free Press.
Hakim, Danny. 2003. "Former employee arrested in university shootings." *Salt Lake Tribune*, May 11. http://www.sltrib.com/2003/May/05112003/nation_w/55881.asp.
Hall, Richard H. (ed.). 1972. *The formal organization*. New York, NY: Basic Books.
Hamilton, Peter. 1987. *The administration of corporate security*. Cambridge, England: ICSA.
Hanson, Julie. 2004. "Politics and policy." *CSO*, January 16.
Haurwitz, Ralph K. M. 2003. "Hackers steal vital data about UT students, staff." *American-Statesman*, March 6. http://www.aup.edu.ph/pipermail/tech/2003-March/000049.html.
Health Canada. 1996. *Laboratory biosafety guidelines*, 2nd ed. http://www.hc-sc.gc.ca/pphb-dgspsp/publicat/lbg-ldmbl-96/.
Howell, Beverly, and Muhammad, Hanif. 2004. "Environmental requirements: Educational institutions accountable." *HTIS Bulletin* 10(6).
Hoye, William P. 2004. "Reducing liability on and off-campus." Eighth Annual Texas Law Conference at the University of North Texas in Denton, Texas, March 1–2.
Integrated Postsecondary Education Data System (IPEDS). 2004. "Employees by faculty status, primary function/occupational activity: Fall 2004." Washington DC: Department of Education, National Center for Education Statistics.
Jordan, Bryce, 2002. *The University of Texas at Dallas, Past, Present and Future Executive Summary*.
Kammen, Daniel M., and Hassenzahl, David M. 1999. *Should we risk it?* Princeton, NJ: Princeton University Press.
Kimberly, John R., Miles, Robert H., and Associates. 1980. *The organizational life cycle*. San Francisco, CA: Jossey-Bass.
Klein, Jan. 2005. "Biosafety officer." Safety Office, The University of Wisconsin at Madison.
Kotter, John P. 1996. *Leading change*. Boston, MA: Harvard Business School Press.
Leedy, Paul D., and Ormrod, Jeanne Ellis. 2001. *Practical research: Planning and design*. Upper Saddle River, NJ: Merrill/Prentice Hall.
Locke, Michelle. 2005. "UC considers using barcodes for cadavers." Associated Press, February 4.
Lofland, John, and Lofland, Lyn H. 1995. *Analyzing social settings: A guide to qualitative observation and analysis*. Belmont, CA: Wadsworth.
Lundegaard, Erik. 2003. "Weathermen's stormy past recounted." *Seattle Times*, October 17, http://seattletimes.nwsource.com/text/2001767647_weather17.html.
MacLeod, Mark. 2004. "Charles Whitman: The Texas Tower sniper." Court TV Crime Library, http://www.crimelibrary.com/notorious_murders/mass/whitman/index_1.html.
Matthews, Karen. 2004. "NYU houses student who lived in library." Associated Press, April 27.
Maxwell, Joseph A. 2005. *Qualitative research design: An interactive approach*. Applied Social Research Methods Series, vol. 41. Thousand Oaks, CA: Sage.
McNabb, David E. 2002. *Research methods in public administration and nonprofit management: Quantitative and qualitative approaches*. New York, NY: M.E. Sharpe.
Meyer, John W., and Rowan, Brian. 1977. "Institutionalized organizations: Formal structure as myth and ceremony." *The American Journal of Sociology* 83(2): 340–363.
Meyer, Marshall W. 1979. *Change in public bureaucracies*. New York, NY: Cambridge University Press.
Michaud, Stephen G., and Aynesworth, Hugh. 2004. "The only living witness: The true story of Ted Bundy." Court TV Crime Library, http://www.crimelibrary.com/criminal_mind/psychology/witness/1.html.
Miller, Gary J. 1998. *Managerial dilemmas: The political economy of hierarchy*. Cambridge, UK: Cambridge University Press.

Moninger, Sara Epstein. 2003. "The flood of '93: Employees deluged with memories a decade later." *FYI: Faculty and Staff News* (The University of Iowa), July 4, 40 (12).

Monroe, Kristen Renwick (ed.). 1991. *The economic approach to politics: A critical reassessment of the theory of rational action*. New York, NY: HarperCollins.

Morgan, Gareth. 1989. *Creative organization theory: A resourcebook*. Newbury Park, CA: Sage.

National Institute of Justice. 2003. (Science and Technology, Fiscal Year, 2004, Solicitation, Funding Resource.) Sl000639. Retrieved on September 25, 2012. https://ncjrs.gov/txtfiles1/nij/S100639.txt.

National Institutes of Health. 2004. "A short history of the National Institute of Health," http://history.nih.gov/exhibits/history/docs/page_02.html.

National Research Council, Building Research Board. 1991. *Uses of risk analysis to achieve balanced safety in building design and operation*. Washington, DC: National Academy Press.

Nolan, Dennis P. 1996. *Handbook of fire and explosion protection engineering principles for oil, gas, chemical and related facilities*. Westwood, NJ: Noyes Publications.

Nuclear Regulatory Commission. 2004. "Our history," http://www.nrc.gov/who-we-are/history.html.

Olswang, Steven J., and Lee, Barbara A. 1984. *Faculty freedoms and institutional accountability: interactions and conflicts*. ASHE-ERIC Higher Education Research Reports 1984, Report 5, Washington, DC.

Oklahoma State University. 1983. "Oklahoma State University policy and procedures." Retrieved on October 24, 2004.

Oklahoma State University. 2012. "Environmental health and safety." Retrieved on July 5, 2012, from http://ehs.okstate.edu/.

Ottley, Ted. 2004. "Ted Kaczynski: The unabomber." Court TV Crime Library, http://www.crimelibrary.com/terrorists_spies/terrorists/kaczynski/1.html.

Pappalardo, Joe. 2009. "Paranoid by design." *Popular Mechanics*, May 2009.

Perez-Pena, Richard. 2004. "Utility could have halted '03 blackout, panel says." *New York Times*, April 6, 2004, http://www.nytimes.com/2004/04/06/national/06BLAC.html?ex=1082258429&ei=1&en=4ec3fa4422e6572c.

Perrow, Charles. 1986. *Complex organizations: A critical essay*, 3rd ed. New York, NY: Random House.

Perrow, Charles. 1999. *Normal accidents: Living with high risk technologies*. Princeton, NJ: Princeton University Press.

"Product of the Month." 2009. *Today's Facility Manager* 21(8).

Purpura, Philip P. 1989. *Modern security and loss prevention management*. Stoneham, MA: Butterworths.

Purtell, Richard. 2009. "Securing office buildings for special events." *Today's Facility Manager* 21(3).

Ramshaw, Emily. 2007. "A&M bioagent workers infected." *Dallas Morning News*. Retrieved on June 27, 2007.

Rankin, Adam. 2004. "AWOL Chinese diplomats ran LANL checkpoint." *Albuquerque Journal*, April 27, http://www.abqjournal.com/scitech/166663science04-27-04.htm.

Raytheon. 2012. "Intelligence Information Systems (IIS) System Safety Program Plan (SSPP) Garland Site," July 20, 2011.

Rhodes, Richard. 1986. *The making of the atomic bomb*. New York, NY: Simon & Schuster.

Schwartz, Ruth. 1987. "The administration and organization of private and public multicampus university libraries: A study of selected cases," AAT 8724091, Columbia University.

"Security lapses found at CDC bioterror lab in Atlanta." *USA Today*. Retrieved on July 5, 2012, from http://www.usatoday.com/news/nation/story/2012-06-27/cdc-lab-security/55870990/1.

Simon, Herbert A. 1997. *Administrative behavior*, 4th ed. New York, NY: The Free Press.

Slavin, Barbara. 2004. "Officials: U.S. 'outed' Iran's spies in 1997." *USA Today*. Retrieved on March 30, 2004, from http://www.usatoday.com/news/washington/2004-03-29-sapphire-usat_x.htm.

Southern California Edison. 2008. "Environmental Impact Analysis and Mitigation Measures, Tehachapi Renewable Transmission Project, Section 4.8.3.1 Regulatory Definitions." Retrieved on February 4, 2008, from http://www.sce.com/nrc/trtp/PEA/4.08_HazMat.htm.

State of Texas. 2008. "Texas Administrative Code, Title 1—Administration, Part 10—Department of Information Resources, Chapter 202 Information Security Standards, Subsection C—Security Standards for Institutions of Higher Education, Rule §202.72, Managing Security Risks." Retrieved on February 4, 2008, from http://info.sos.state.tx.us/pls/pub/readtac$ext.TacPage?sl=R&app=9&p_dir=&p_rloc=&p_tloc=&p_ploc=&pg=1&p_tac=&ti=1&pt=10&ch=202&rl=72.

SUNY-Albany, 2012. Environmental Health and Safety. Retrieved on January 11, 2012. http://www.albany.edu/.

SUNY-Binghamton, 2004. Division of Administration, Organizational Chart. http://www.binghamton.edu/.

Texas A&M University–Corpus Christi. 2004. "Chemical safety." Retrieved on October 25, 2004, from http://safety.tamucc.edu/tamucc/chemical.html.

Texas A&M University–Corpus Christi. 2012. "Texas A&M University—Corpus Christi—Safety procedures and guidelines." Retrieved on July 5, 2012, from http://safety.tamucc.edu/uploads/Site/safety.pdf.

Texas Department of Insurance. 2004. "State fire marshal." Retrieved on October 24, 2004, from http://www.tdi.state.tx.us/fire/fmsti.html.

Thompson, Fay. 1991. "Managing Academe's Hazardous Materials." *Planning for Higher Education* 19(2): 1–5.

Tulane University Health Sciences Center. 2004. "Hazardous Materials and Waste. OEHS policies." Retrieved on October 24, 2004.

Tulane University Health Sciences Center. 2012. "Office of Environmental Health and Safety (OEHS)." Retrieved on July 5, 2012, from http://tulane.edu/oehs/.

Twohey, Megan. 2003. "National security restrictions crimp university research." *Milwaukee Journal Sentinel*, December 1.

Union Carbide. 2007. "Bhopal Information Center: Chronology." Retrieved February 23, 2008, from http://www.bhopal.com/chrono.htm.

University at Albany, State University of New York. 2012. "Environmental health and safety." Retrieved on January 11, 2012, from http://www.albany.edu/.

University of California–Berkeley. 2005. "About the Valley Life Sciences Building." Retrieved on February 23, 2005, from http://ib.berkeley.edu/vlsb/about.html.

University of California–Riverside. 2012. "Environmental health and safety." Retrieved on January 12, 2012, from http://ehs.ucr.edu/.

University of California–Santa Barbara. 2012. "Environmental health and safety." Retrieved on July 5, 2012, from http://www.ehs.ucsb.edu/.

University of California–Santa Cruz. 2004. "Business and administrative services." Retrieved on October 25, 2004, from http://www.ucsc.edu/public/.

University of California–Santa Cruz. 2012. "Environmental health and safety." Retrieved on January 12, 2012, http://ehs.ucsc.edu/.

University of Maine. 2004. "Environmental health and safety manual." Retrieved on October 25, 2004.

University of Maryland–Baltimore County. 2004. "Police crisis response," http://www.umbc.edu/police/crisis_response/hazardous_materials.htm.

University of Maryland–Baltimore County. 2012. "Environmental health and safety." Retrieved on January 12, 2012, from http://www.ehs.umaryland.edu/.

University of New Brunswick, Office of Campus Safety. 2004. "Laboratory safety." Retrieved on October 24, 2004, from http://www.unb.ca/safety/labsafety.html.

University of North Carolina–Greensboro. 2012. "Environmental health and safety." Retrieved on January 11, 2012, from http://www.uncg.edu/sft/.

University of Texas at Arlington. 2004. "Chemical and biological laboratory safety." Retrieved on October 24, 2004, from www.uta.edu/ehsafety/safetyplan/11biochm.pdf.

University of Texas at Arlington. 2012. "Administration and campus operations: Environmental health and safety." Retrieved on July 5, 2012, from http://www.uta.edu/campus-ops/ehs/.

University of Texas at Austin. 2004. "Vice president for employee and campus services." Retrieved on October 24, 2004, from http://www.utexas.edu/.

University of Texas at Austin. 2012. "Environmental health and safety." Retrieved on January 12, 2012, from http://www.utexas.edu/safety.

University of Texas System. 2004. *Accountability and Performance Report*. Austin, Texas.

University of Wisconsin–Madison. 2012. "Environmental health and safety department." Retrieved on January 12, 2012, from http://www.wisc.edu/.

U.S. Air Force. 1997. Radioactive Material Management, September 1997 - TI#14098. http://www.p2pays.org/ref/07/06063.htm.

U.S. Congress. 1988. Clinical Laboratory Improvement Amendment. Public Law 100-578, October 31.

U.S. Congress. 2002. Homeland Security Act 2002. H.R. Bill 5005, January 23.

Valcik, Nicolas. 2006. *Regulating the use of biological hazardous materials in universities: Complying with the new federal guidelines*. Lewiston, NY: Edwin Mellen Press.

Vergano, Dan, and Sternberg, Steve. 2004. "Anthrax slip-ups raise fears about planned bio-labs." *USAToday*, October 14.

Villano, Matt. 2009. "Eureka!" *Campus Technology* 22(8).

Vinas, Maria Jose. 2008. "Animal-rights militants torch UCLA van." *Chronicle of Higher Education* 55(42).

Warren, Carol A. B., and Karner, Tracy X. 2005. *Discovering qualitative methods: Field research, interviews, and analysis*. Los Angeles, CA: Roxbury.

Webb, Eugene J., Campbell, Donald T., Schwartz, Richard D., and Sechrest, Lee. 1973. *Unobtrusive measures: Nonreactive research in the social sciences*. Chicago, IL: Rand McNally.

"With Some Strings Attached." 2003. *CSO*, December 9.

Wright, Matt. 2004. "Report: Los Alamos violated 7 safety rules." *The Daily Texan*, July 25www.dailytexanonline.com.

Yale University. 2000. "Guidelines for Safe Laboratory Practices, Department of Chemistry." Retrieved on October 25, 2004, from http://216.109.117.135/search/cache?p=biological+laboratory+guidelines&xargs=0&pstart=1&b=21&u=www.chem.yale.edu/safety/SafetyManual00.pdf&w=biological+laboratory+guidelines&d=842C239F20&c=482&yc=25761&icp=1.

Youngers, Jane A., and Norris, Julie T. 2004. *Regulations and compliance: A compendium of regulations and certifications applicable to sponsored programs*. Washington, DC: National Council of University Research Administrators (NCURA).

Index

K

L

M

N

O

P